Universitext

Universitext

Universitext is a series of textbooks that presents material from a wide variety of mathematical disciplines at master's level and beyond. The books, often well class-tested by their author, may have an informal, personal even experimental approach to their subject matter. Some of the most successful and established books in the series have evolved through several editions, always following the evolution of teaching curricula, to very polished texts.

Thus as research topics trickle down into graduate-level teaching, first textbooks written for new, cutting-edge courses may make their way into *Universitext*.

More information about this series at http://www.springer.com/series/223

Joel H. Shapiro

A Fixed-Point Farrago

Joel H. Shapiro
Portland State University
Portland, OR, USA

ISSN 0172-5939   ISSN 2191-6675   (electronic)
Universitext
ISBN 978-3-319-27976-3   ISBN 978-3-319-27978-7   (eBook)
DOI 10.1007/978-3-319-27978-7

Library of Congress Control Number: 2015958889

Printed on acid-free paper

This Springer imprint is published by Springer Nature
The registered company is Springer International Publishing AG Switzerland

For Lilah

Farrago—noun

An assortment or a medley; a conglomeration.

thefreedictionary.com

Preface

Fixed points show up everywhere in mathematics. This book provides an introduction to some of the subject's best-known theorems and some of their most important applications, emphasizing throughout their interaction with topics in the analysis familiar to students of mathematics. The level of exposition increases slowly, requiring at first some undergraduate-level proficiency, then gradually increasing to the kind of sophistication one might expect from a graduate student. Appendices at the back of the book provide introduction to (or reminder of) some of the prerequisite material. To encourage active participation, exercises are integrated into the text. Thus I hope readers will find the book reasonably self-contained and useable either on its own or as a supplement to standard courses in the mathematics curriculum.

The material is split into four parts, the first of which introduces the Banach Contraction Mapping Principle and the Brouwer Fixed-Point theorem, along with a selection of interesting applications: Newton's method, initial value problems, and positive matrices (e.g., the Google matrix). Brouwer's theorem is proved in dimension two via Sperner's lemma, and Banach's principle is proved in generality. Included also is a lesser known fixed-point theorem due to Knaster and Tarski—an easy-to-prove result about functions taking sets to sets that makes short work of the Schröder–Bernstein theorem and plays an important role in a later chapter on paradoxical decompositions.

Part II focuses on Brouwer's theorem, featuring an analysis-based proof of the general result, and John Nash's application of this result to the existence of Nash equilibrium. Brouwer's theorem leads to Kakutani's theorem on set-valued maps, upon which rests Nash's remarkable "one-page" proof of his famous theorem. A brief introduction to game theory motivates the exposition of Nash's results.

The material of these first two parts should be accessible to undergraduates whose background includes the standard junior–senior-level courses in linear algebra and analysis taught at American colleges, which hopefully provides some familiarity with basic set theory and metric spaces.

Part III applies Brouwer's theorem to spaces of infinite dimension, where it provides an important step in the proof of the Schauder Fixed-Point theorem. Schauder's theorem leads to both Peano's existence theorem for initial value problems and Lomonosov's spectacular theorem on invariant subspaces for linear operators on Banach spaces. For this segment the reader needs only some experience with the basics of Hilbert and Banach spaces.

The fourth and final part of the book rests on the work of Markov, Kakutani, and Ryll-Nardzewski concerning fixed points for families of affine maps. These results lead to the existence of measures—both finitely and countably additive—that are invariant under various groups of transformations. In the finitely additive case, this leads to the concepts of invariant means and "paradoxical decompositions," especially the Banach–Tarski paradox. The countably additive case leads to the existence of Haar measure on compact topological groups. This part of the book gets into notions of duality and weak-star topologies, with the necessary prerequisites developed from scratch—but only within the narrow context in which they are used. The result is a gentle introduction to abstract duality which suffices for our purposes, and hopefully encourages the reader to appreciate this way of thinking.

Much of the material presented here originated in lectures given during the academic years 2012–2013 by participants in the Analysis Seminar at Portland State University. I am particularly indebted to John Erdman, who organized the seminar for many years and who introduced us to the Knaster–Tarski theorem; to Steve Silverman who lectured on the work of Markov and Kakutani; to Mau Nam Nguyen, Blake Rector, and Jim Rulla for their talks on set-valued analysis; to Steve Bleiler and Cody Fuller for their lectures on game theory; and to all the seminar participants whose thoughtful questions and comments contributed greatly to my appreciation of the subject.

Sheldon Axler encouraged me to turn my lecture notes into a book and suggested that Sperner's lemma might have a place in it. Paul Bourdon contributed many insightful comments on initial versions of manuscript and cleaned up several of my more cumbersome arguments. The Fariborz Maseeh Department of Mathematics and Statistics at Portland State University provided office space and technical assistance, and Michigan State University—my employer in a former life—provided invaluable electronic access to its library.

Above all, I owe a deep debt of gratitude to my wife Jane, without whose understanding and encouragement this project could never have reached completion.

Portland, OR, USA   Joel H. Shapiro
August 2015

Contents

Part II From Brouwer to Nash

Part III Beyond Brouwer: Dimension $= \infty$

Appendices

These first three chapters explain the idea of "fixed point" and, in a manner accessible to readers with a background in undergraduate mathematics, motivate its importance.

Chapter 1
From Newton to Google

Overview. After setting out the definition of "fixed point" we'll give examples of their role in finding solutions: to equations (Newton's Method), to initial-value problems, and to the problem of ranking internet web pages. After this we'll show how the notion of fixed point arises in set theory, where it provides an easy proof of the Schröder–Bernstein theorem. We'll introduce the famous Brouwer Fixed-Point Theorem, show how it applies to the study of matrices with positive entries, and discuss the application of these results to the internet page-ranking problem.

Prerequisites. Calculus (continuity and fundamental theorem of integral calculus), differential equations (initial-value problems), basic linear algebra, some familiarity with sets and the operations on them.

1.1 What Is a Fixed Point?

Definition (Fixed Point). Suppose f is a map that takes a set S into itself. A *fixed point* of f is just a point $x \in S$ with $f(x) = x$.

A map f can have many fixed points (example: the identity map on a set with many elements) or no fixed points (example: the mapping of "translation-by-one," $x \rightarrow x + 1$ on the real line).

Exercise 1.1. The fixed points of a function mapping a real interval into itself can be visualized as the x-coordinates of the points at which the function's graph intersects the line $y = x$. Use this idea to help in determining the fixed points possessed by each of the functions $f : \mathbb{R} \rightarrow \mathbb{R}$ defined below.

(a) $f(x) = \sin x$

(b) $f(x) = x + \sin x$

(c) $f(x) = 2 \sin x$

© Springer International Publishing Switzerland 2016
J.H. Shapiro, *A Fixed-Point Farrago*, Universitext,
DOI 10.1007/978-3-319-27978-7_1

1.2 Example: Newton's Method

Suppose for simplicity that f is a differentiable function $\mathbb{R} \to \mathbb{R}$, with derivative f' continuous and never vanishing on $\mathbb{R}$. Consider for f its "Newton function" F, defined by

$$F(x) = x - \frac{f(x)}{f'(x)} \qquad (x \in \mathbb{R}). \tag{1.1}$$

One can think of $F(x)$ as the horizontal coordinate of the point at which the line tangent to the graph of f at the point $(x, f(x))$ intersects the horizontal axis. Since f' doesn't vanish, F is a continuous mapping taking $\mathbb{R}$ into itself. The roots of f (those points $x \in \mathbb{R}$ such that $f(x) = 0$) are precisely the fixed points of F.

Newton's method involves iterating the Newton function in the hope of generating approximations to the roots of f. One starts with an initial guess x_0, sets $x_1 = F(x_0), x_2 = F(x_1) \ldots$, and hopes that the resulting sequence of "Newton iterates" converges to a fixed point of F. Geometrically it seems clear that if the Newton iterate sequence converges then it must converge to a root of f. We'll see later, as a consequence of something far more general (Proposition 3.3, page 28), that this indeed the case.

1.3 Example: Initial-Value Problems

From a continuous function $f \colon \mathbb{R}^2 \to \mathbb{R}$ and a point $(x_0, y_0) \in \mathbb{R}^2$ we can create an *initial-value problem*

$$y' = f(x, y), \quad y(x_0) = y_0. \tag{IVP}$$

Geometrically, (IVP) asks for a differentiable function y whose graph is a smooth "solution curve" in the plane that has the following properties:

(a)   The curve passes through the point (x_0, y_0), and
(b)   at each of its points (x, y) the curve has slope $f(x, y)$.

As a first attempt to solve the differential equation $y' = f(x, y)$ one might try integrating both sides with respect to x. If by "integrate both sides" we mean "take the definite integral from x_0 to x," then there results the *integral equation*

$$y(x) = y_0 + \int_{t=x_0}^{x} f(t, y(t)) \, dt \tag{IE}$$

which is implied by (IVP) in the sense that each function y satisfying (IVP) for some interval of x's containing x_0, also satisfies (IE) for that same interval.

Conversely, suppose $y \in C(\mathbb{R})$ satisfies (IE) on some open interval I. Fix $x \in I$. Then for $h \in \mathbb{R} \backslash \{0\}$ small enough that $x + h \in I$, the Mean-Value Theorem of integral calculus provides a point ξ between x and $x + h$ such that

$$\frac{y(x+h) - y(x)}{h} = \frac{1}{h}\int_x^{x+h} f(t, y(t))\, dt = f(\xi, y(\xi)).$$

Thanks to the continuity of f, as $h \to 0$ the expression on the right, and therefore the difference quotient on the left, converges to $f(x, y(x))$. Thus y is differentiable at x and $y'(x) = f(x, y(x))$, i.e., the function y satisfies the differential equation in (IVP) on the interval I. That it satisfies the initial condition is trivial.

Conclusion: (IVP) $\equiv$ (IE).

To make the connection with fixed points, let $C(\mathbb{R})$ denote the vector space of continuous, real-valued functions on $\mathbb{R}$, and consider the *integral transform* $T: C(\mathbb{R}) \to C(\mathbb{R})$ defined by

$$(Ty)(x) = y_0 + \int_{t=x_0}^x f(t, y(t))\, dt \qquad (x \in \mathbb{R}). \tag{1.2}$$

Thus equation (IE) can thus be rewritten $Ty = y$, so to say $y \in C(\mathbb{R})$ satisfies (IVP) turns out to be the same as saying: *y is a fixed point of the mapping T*. In Chap. 3 we'll discuss the existence and uniqueness of such fixed points.

1.4 Example: The Internet

At each instant of time the publicly accessible Internet consists of a collection of N web pages[1] each of which can have links coming in from, and going out to, other pages. To be effective, search engines such as Google must seek to determine the importance of each individual page. Here's a first attempt to do this.

To each web page P_i ($1 \le i \le N$) we'll assign a non-negative real number $\mathrm{imp}(P_i)$ that measures the "importance" of that page. The page P_i will derive its importance from all the pages that link into it: if P_j has a total of λ_j outgoing links then we decree that it bestow importance of $\mathrm{imp}(P_j)/\lambda_j$ to each of the pages into which it has links. In other words, if we think of P_j as having $\mathrm{imp}(P_j)$ "votes" then our rule is that it must distribute these votes evenly among the pages into which it links. The importance of a given web page is then defined to be the sum of the importances it receives from each of the web pages that link to it (self-links are allowed).

This definition of "importance" for a web page may seem at first glance to be circular, but it's not! To make matters precise, define L_i to be the set consisting of all indices j for which P_j links into P_i. Then $\mathrm{imp}(P_i)$ is given by the equation

$$\mathrm{imp}(P_i) = \sum_{j \in L_i} \frac{1}{\lambda_j} \mathrm{imp}(P_j) \qquad (1 \le i \le N) \tag{1.3}$$

---

[1] According to www.worldwidewebsize.com, N was $\ge 4.74 \times 10^9$ on August 4, 2015.

This is a set of N linear equations in the N unknowns $\operatorname{imp}(P_i)$; for these equations to provide a reasonable ranking of web pages there needs to be a solution not identically zero, all coordinates of which are non-negative.

To see how fixed points enter into this discussion, let v to be the column vector with $\operatorname{imp}(P_i)$ in the i-th position, and define the "hyperlink matrix" H to be the $N \times N$ matrix whose j-th column has $1/\lambda_j$ in the i-th entry if P_j links to P_i, and zero otherwise. With these definitions Eq. (1.3) can be rewritten in matrix form

$$v = Hv,$$

so the "importance vector" v we seek is a fixed point of the transformation that H induces on the set of vectors in $\mathbb{R}^N \backslash \{0\}$, all of whose coordinates are non-negative.

In the language of linear algebra: we demand that 1 be an eigenvalue of H and v be a corresponding eigenvector *with non-negative entries*. We'll see in Sect. 1.7 that such a vector actually exists. In Sect. 1.8 we'll also take up the crucial question of uniqueness (at least up to positive scalar multiples); to be effective our method needs to produce an *unequivocal* ranking of web pages.

Mini-example. Consider the fictional mini-internet pictured in Fig. 1.1 below which consists of six web pages, the label on each link denoting the proportion of the donor page's importance being granted to the recipient page.

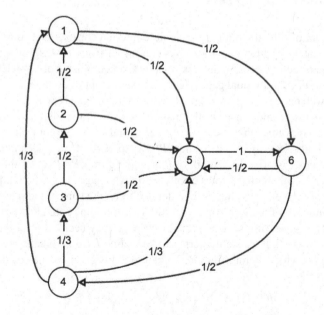

Fig. 1.1 An imaginary six-page internet

The hyperlink matrix for our mini-web is

$$H_0 = \begin{bmatrix} 0 & \frac{1}{2} & 0 & \frac{1}{3} & 0 & 0 \\ 0 & 0 & \frac{1}{2} & 0 & 0 & 0 \\ 0 & 0 & 0 & \frac{1}{3} & 0 & 0 \\ 0 & 0 & 0 & 0 & 0 & \frac{1}{2} \\ \frac{1}{2} & \frac{1}{2} & \frac{1}{2} & \frac{1}{3} & 0 & \frac{1}{2} \\ \frac{1}{2} & 0 & 0 & 0 & 1 & 0 \end{bmatrix}$$

for which your favorite matrix calculation program will verify that 1 is an eigen-value, so the equation $H_0 v = v$ does have a non-zero solution in $\mathbb{R}^6$. Furthermore, the calculation will show that this solution is unique up to scalar multiples, and all of its entries have the same sign. When normalized to have positive entries and Euclidean norm 1, this vector is, to two significant digits,

$$v = [0.14, \ 0.057, \ 0.11, \ 0.34, \ 0.61, \ 0.69]^t$$

where the superscript "t" denotes "transpose." This gives the following ranking, from most to least important: $(P_6, \ P_5, \ P_4, \ P_1, \ P_3, \ P_2)$.

This example illustrates a critical fact about the process of ranking pages: the page with the most incoming links need not be the most important! In particular, Page 6, with only two incoming links, is more important than Page 5, which has five incoming links. Similarly, Page 4, with just one incoming link, is more important than Page 1, which has two such links (Exercise: Can you explain in just a few words what's making this happen?).

Exercise 1.2. "Importance vectors" need not be unique. Consider the following mini-web, still with six pages. but now with links that look like this:

$$1 \to 2 \to 3 \to 1 \quad \text{and} \quad 4 \to 5 \to 6 \to 4.$$

Write out the hyperlink matrix for this mini-web and show that it has several independent importance vectors (some of which contain zeros).

The hyperlink matrix H for the full internet, although huge, consists mostly of zeros; each web page links to a relatively tiny number of others.[2] Furthermore each column of H corresponding to a page with outlinks will sum to 1, but each column corresponding to a page with *no* outlinks is identically zero. This latter kind of page (a particularly annoying one) is called a "dangling node." Were the internet to have no such pages, its hyperlink matrix would be *stochastic*: each entry non-negative and each column summing to one. This is the case for the six-page example worked out above, as well as the mini-web of Exercise 1.2.

---

[2] According to [12], on average somewhere in the hundreds.

Stochastic matrices have particularly nice properties; we'll show in Sect. 1.7 that their associated linear transformations possess nontrivial fixed points (i.e., 1 is an eigenvalue) *with non-negative entries*. Thus, were our internet to have no dangling nodes, the hyperlink matrix would have a fixed point that would provide a ranking of websites. For this reason we'd like to find a modification of H that achieves "stochasticity" without compromising the intuition behind our definition of "importance."

One way to do this is to think of a dangling node as, rather than linking to *no* other web pages, actually linking to *every* web page (including itself), contributing $1/N$ of its importance to every web page. This models the behavior of a web surfer who, stuck at a page with no outlinks, decides to skip directly to a random page, thus establishing to that page a "link" of weight $1/N$. Our new $N \times N$ hyperlink matrix, call it H_1, is now stochastic; the columns previously identically zero are now identically $1/N$. We'll return to this matrix in Sect. 1.7.

1.5 Example: The Schröder–Bernstein Theorem

The famous Schröder–Bernstein theorem of set theory asserts:

> If X and Y are sets for which there is a one-to-one mapping taking X into Y and a one-to-one mapping taking Y into X, then there is a one-to-one mapping taking X *onto* Y.

This result follows from a more general one depicted in Fig. 1.2 below:

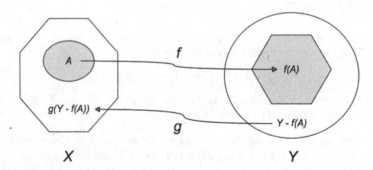

Fig. 1.2 The Banach Mapping Theorem

Theorem 1.1 (The Banach Mapping Theorem). *Given sets X and Y and functions $f: X \to Y$ and $g: Y \to X$, there is a subset A of X whose complement is the g-image of the complement of $f(A)$.*

To see how the Banach Mapping Theorem implies the Schröder–Bernstein Theorem, suppose in the statement of the Banach theorem that the maps f and g are one-to-one. Then the map $h: X \to Y$ defined by setting $h = f$ on A and $h = g^{-1}$ on $X \backslash A$ is the one promised by Schröder–Bernstein. $\qquad\square$

The Banach Mapping Theorem is, in fact, a fixed-point theorem! Its conclusion is that there is a subset A of X for which $X \backslash A = g(Y \backslash f(A))$. This equation is equivalent, upon complementing both sides in X, to

$$A = X \backslash g(Y \backslash f(A)). \tag{1.4}$$

For a set S, let's write $\mathscr{P}(S)$ for the collection of all subsets of S. Define the function $\Phi: \mathscr{P}(X) \to \mathscr{P}(X)$ by

$$\Phi(E) = X \backslash g(Y \backslash f(E)) \qquad (E \in \mathscr{P}(X)).$$

With these definitions, Eq. (1.4) asserts that the set A is a fixed point of Φ.

That such a fixed point exists is not difficult to prove. The mapping Φ defined above is best understood as the composition of four simple set-mappings:

$$\mathscr{P}(X) \xrightarrow{f} \mathscr{P}(Y) \xrightarrow{C_Y} \mathscr{P}(Y) \xrightarrow{g} \mathscr{P}(X) \xrightarrow{C_X} \mathscr{P}(X)$$

where C_X denotes "complement in X" and similarly for C_Y, while f and g now denote the "set functions" induced in the obvious way by the original "point functions" f and g. Since f and g preserve set-containment (i.e., $E \subset F \implies f(E) \subset f(F)$) while C_X and C_Y reverse it, the composite mapping Φ *preserves* set-containment.

With these observations, the theorems of Banach and Schröder–Bernstein follow from:

Theorem 1.2 (The Knaster–Tarski Theorem). *If X is a set and $\Phi: \mathscr{P}(X) \to \mathscr{P}(X)$ is a mapping that preserves set-containment, then Φ has a fixed point.*

Proof. Let $\mathscr{E}$ be the collection of subsets E of X for which $E \subset \Phi(E)$. Since $\mathscr{E}$ contains the empty subset of X, it is nonempty. Let A denote the union of all the sets in $\mathscr{E}$.

Claim. $\Phi(A) = A$.

Proof of Claim. Suppose $E \in \mathscr{E}$. Then $E \subset \Phi(E)$ by the definition of $\mathscr{E}$, and $E \subset A$ by the definition of A. Thus $\Phi(E) \subset \Phi(A)$ by the containment-preserving nature of Φ, hence $E \subset \Phi(A)$. Consequently $A \subset \Phi(A)$, whereupon $\Phi(A) \subset \Phi(\Phi(A))$, which places $\Phi(A)$ in $\mathscr{E}$. Conclusion: $\Phi(A) \subset A$, hence $\Phi(A) = A$. $\qquad\square$

1.6 The Brouwer Fixed-Point Theorem

The most easily stated—and deepest—of the fixed-point theorems we'll discuss in this book was proved in 1912 by the Dutch mathematician L.E.J. Brouwer. Its initial setting is the closed unit ball B of Euclidean space $\mathbb{R}^N$.

Theorem 1.3 (The Brouwer Fixed-Point Theorem). *Every continuous mapping of B into itself has a fixed point.*

It's easy to see that the result remains true if B is replaced by a homeomorphic image (i.e., a set $G = f(B)$ where f is continuous, one-to-one, with $f^{-1}: G \to B$ also continuous).

For $N = 1$ the proof of Brouwer's Theorem is straightforward. In this case f is a continuous function mapping the real interval $[-1, 1]$ into itself. We may suppose f doesn't fix either endpoint (otherwise we're done), so $f(-1) > -1$ and $f(1) < 1$. In other words, the value of the continuous function $g(x) = f(x) - x$ is positive at $x = -1$ and negative at $x = 1$. By the Intermediate Value Theorem, g must take the value zero at some point of the interval $(-1, 1)$; that point is a fixed point for f.

The proof for $N > 1$ is much more difficult, and there are many different versions. For $N = 2$ we'll prove in the next chapter a famous combinatorial lemma due to Sperner which yields Brouwer's Theorem for that case,[3] and in Chap. 4 we'll prove the full result using methods of "advanced calculus."

The Brouwer Theorem for convex sets. To say a subset of $\mathbb{R}^N$, or more generally of a vector space over the real field, is *convex* means that if two points belong to the set, then so does the entire line segment joining those points. More precisely:

Definition 1.4. To say a subset C of a real vector space is *convex* means that: whenever x and y belong to C then so does $tx + (1 - t)y$ for every real t with $0 \le t \le 1$.

In the course of proving the Brouwer Theorem for $\mathbb{R}^N$ we'll develop enough machinery to obtain it for all closed, bounded convex sets therein (Theorem 4.5). Officially:

Theorem 1.5 (The "Convex" Brouwer Fixed-Point Theorem). *Suppose N is a positive integer and C is a closed, bounded, convex subset of $\mathbb{R}^N$. Then every continuous mapping taking C into itself has a fixed point.*

1.7 Application: Stochastic Matrices

Recall from Sect. 1.4 that a *stochastic matrix* is a square matrix that is *non-negative* (all entries ≥ 0), all of whose columns sum to 1. In that section (see page 8) we offered as an example the modified internet hyperlink matrix H_1: the original

---

[3] Our version of Sperner's Lemma generalizes to dimension $N > 2$, where it also implies Brouwer's Theorem. However we will not pursue this direction.

internet hyperlink matrix with the zero-columns replaced by "$1/N$-columns." H_1 is an $N \times N$ stochastic matrix with N on the order of several billion. Our proposed method for ranking internet websites depended on finding a vector $v \in \mathbb{R}^N \backslash \{0\}$ with non-negative entries such that $H_1 v = v$. Now there's no secret that $H_1 v = v$ for *some* vector $v \in \mathbb{R}^N \backslash \{0\}$, i.e., that 1 is an eigenvalue for H_1; in fact this is true of *every* stochastic matrix.

To see why, let e denote the (column) vector in $\mathbb{R}^N$, all of whose entries are 1. Let A be an $N \times N$ stochastic matrix. Since all the columns of A sum to 1 we have $A^t e = e$, where the superscript "t" denotes "transpose." Thus 1 is an eigenvalue of A^t. Since each square matrix has the same eigenvalues as its transpose (the determinant of a square matrix is the same as that of its transpose, hence both matrix and transpose have the same characteristic polynomial) we see that 1 is an eigenvalue of A, i.e., there exists $x \in \mathbb{R}^N \backslash \{0\}$ with $Ax = x$. However, to be meaningful for the internet our eigenvector must have all coordinates non-negative and—up to positive scalar multiples—be unique.

Uniqueness is a special problem. For example, the $N \times N$ identity matrix is stochastic, but (if $N > 1$) has lots of essentially different non-negative fixed points. We'll return to this question in the next chapter. Right now let's see how the Brouwer Fixed-Point Theorem proves that, questions of uniqueness aside:

Theorem 1.6. *Every stochastic matrix has a fixed point, all of whose entries are non-negative, and at least one of which is positive.*

In particular, the modified hyperlink matrix H_1 of Sect. 1.4 has a fixed point that produces at least one ranking of web pages.

For the proof of Theorem 1.6 we'll view $\mathbb{R}^N$ as a space of column vectors, but with distances measured in the metric arising from the "one-norm:"

$$\|x\|_1 = |\xi_1| + |\xi_2| + \cdots + |\xi_N| \qquad (x \in \mathbb{R}^N), \qquad (1.5)$$

where ξ_j is the j-th coordinate of the vector x.

Exercise 1.3. Check that $\| \cdot \|_1$ is a norm[4] on $\mathbb{R}^N$, and that

$$\frac{1}{\sqrt{N}} \|x\| \le \|x\|_1 \le \sqrt{N} \|x\| \qquad (x \in \mathbb{R}^N)$$

where $\|x\| = (\sum_j \xi_j^2)^{1/2}$, the Euclidean norm of the vector x. Show that this implies that the distance d_1 defined on $\mathbb{R}^N$ by $d_1(x, y) = \|x - y\|_1$ is *equivalent* to the one induced by the Euclidean norm, in that both distances give rise to the same convergent sequences.

Definition 1.7 (Standard Simplex). The *standard N-simplex* Π_N is the set of non-negative vectors in the closed $\| \cdot \|_1$-unit "sphere" of $\mathbb{R}^N$, i.e.,

$$\Pi_N = \{x \in \mathbb{R}_+^N : \|x\|_1 = 1\},$$

where $\mathbb{R}_+^N$ denotes the set of vectors in $\mathbb{R}^N$ with all coordinates non-negative.

---

[4] For the definition of "norm" see Appendix C.2, page 194.

For example, Π_2 is the line segment in $\mathbb{R}^2$ joining the points $(0,1)$ and $(1,0)$, while Π_3 is the triangle in $\mathbb{R}^3$ with vertices $(1,0,0)$, $(0,1,0)$, and $(0,0,1)$. In general Π_N is the *convex hull* of the standard unit vector basis in $\mathbb{R}^N$, the smallest convex subset of $\mathbb{R}^N$ that contains those vectors (Proposition C.4 of Appendix C).

Exercise 1.4. Show that Π_N is closed and bounded in $\mathbb{R}^N$, hence compact.

Proof of Theorem 1.6. Our goal is to show that A (more accurately: the linear transformation that A induces on $\mathbb{R}^N$) has a fixed point in Π_N. For this it's enough to show that $A(\Pi_N) \subset \Pi_N$, after which the continuity of A (see Exercises 1.5 and 1.6 below), the compactness of Π_N (Exercise 1.4 above), and Theorem 1.5 (the "Convex" Brouwer Fixed-Point Theorem) will combine to produce the desired fixed point.

To see that the matrix A takes Π_N into itself, let's denote by $a_{i,j}$ the element of A in row i and column j. Fix $x \in \Pi_N$, and let ξ_j denote its j-th coordinate. Then, since all matrix elements and coordinates are non-negative:

$$\|Ax\|_1 = \sum_{i=1}^{N} \left(\sum_{j=1}^{N} a_{i,j}\xi_j \right) = \sum_{j=1}^{N} \left(\sum_{i=1}^{N} a_{i,j} \right) \xi_j = \sum_{j=1}^{N} \xi_j = 1,$$

the third equality reflecting the fact that, to its left, the sum in parentheses is the j-th column-sum of A, which by "stochasticity" equals 1. Thus $x \in \Pi_N$, as desired. $\square$

Regarding the Internet. Theorem 1.6 establishes that the modified hyperlink matrix H_1 has fixed points in Π_N, and so yields (possibly many) rankings of web pages. This issue of non-uniqueness arose in Exercise 1.2; we'll resolve it in the next section.

Exercise 1.5. Modify the argument above to show that if A is a stochastic matrix then $\|Ax\|_1 \leq \|x\|_1$ for every $x \in \mathbb{R}^N$ (and even for every $x \in \mathbb{C}^N$).

Exercise 1.6. Use the inequality of the previous exercise to show that A, or more accurately the linear transformation A induces on $\mathbb{R}^N$ (and even on $\mathbb{C}^N$), is continuous in the distances induced on $\mathbb{R}^N$ (and even on $\mathbb{C}^N$) by both the one-norm and the more familiar Euclidean norm.

Exercise 1.7. Modify the ideas in the previous two exercises to establish the continuity of the linear transformation induced on $\mathbb{R}^N$ (and even on $\mathbb{C}^N$) by *any* $N \times N$ real matrix.

Exercise 1.8. Theorem 1.6 shows that every stochastic matrix has 1 as an eigenvalue. Use Exercise 1.5 to show that no eigenvalue, real or complex, has larger modulus. In matrix-theory language: Every stochastic matrix has *spectral radius* 1.

Exercise 1.9. Show that the collection of $N \times N$ stochastic matrices is a convex subset of the real vector space of all $N \times N$ matrices.

1.8 Perron's Theorem

To call a real matrix A of any dimensions (square, row, column ...) "non-negative" (written "$A \geq 0$") means that all its entries are non-negative, and to call it "positive" (written "$A > 0$") means that all its entries are strictly positive.[5] Our first result picks up where Theorem 1.6 above left off, and forms the core of Perron's famous 1907 theorem on eigenvalues of positive matrices.

Theorem 1.8. *Every positive square matrix has a positive eigenvalue, to which corresponds a positive eigenvector.*

We'll prove this by modifying the argument used for Theorem 1.6. The difficulty to be overcome is that, without the hypothesis of stochasticity, our matrices need not take the standard N-simplex Π_N into itself. This is easy to fix.

Lemma 1.9. *Suppose A is an $M \times N$ matrix that is > 0. Then $Ax > 0$ for every vector $x \in \mathbb{R}^N_+ \backslash \{0\}$.*

In words: If every entry of A is strictly positive and every entry of $x \in \mathbb{R}^N \backslash \{0\}$ is non-negative, then every entry of Ax is strictly positive.

Proof of Lemma. Suppose $x \in \mathbb{R}^N_+ \backslash \{0\}$. Fix an index j and note that the j-th coordinate of Ax is the dot product of the j-th row of A with the (transpose of the) column vector x. Since the entries of A are all strictly positive, and the entries of x are non-negative and not all zero, this dot product is strictly positive. $\qquad\square$

Proof of Theorem. Suppose A is an $N \times N$ positive matrix. Lemma 1.9 insures that $Ax > 0$ for every $x \in \Pi_N$, hence the equation

$$F(x) = \frac{Ax}{\|Ax\|_1} \qquad (x \in \Pi_N)$$

defines a map F that's continuous on Π_N and takes that simplex into itself. Theorem 1.5 then guarantees for F a fixed point $x_0 \in \Pi_N$. Thus x_0 is a vector with non-negative coordinates, $\|x_0\|_1 = 1$, and $Ax_0 = \lambda x_0$, where $\lambda = \|Ax_0\|_1 > 0$, hence $Ax_0 > 0$ by Lemma 1.9. Conclusion: $x_0 = \lambda^{-1}Ax_0 > 0$. $\qquad\square$

Perron Eigenpairs. Theorem 1.8 guarantees that every positive matrix A has what we might call a *Perron eigenpair* (λ, x): a positive ("Perron") eigenvalue λ that has a positive ("Perron") eigenvector x with $\|x\|_1 = 1$. Now we've seen in Theorem 1.6 that every *stochastic* matrix has a "weak" Perron eigenpair $(1, x)$ ("weak" because some coordinates of x may be zero), and in Exercise 1.8 that for stochastic matrices, no eigenvalue (real or complex) has modulus larger than 1. Our next result derives a stronger conclusion from a weaker hypothesis.

Theorem 1.10 (Perron's Theorem). *Each positive square matrix A possesses exactly one Perron eigenpair. Among all the (possibly complex) eigenvalues of A, the Perron eigenvalue has the largest modulus.*

---

[5] *Warning:* This is not to be confused with the notion of "positive-definite," which is something completely different.

Proof. Suppose A is a positive $N \times N$ matrix and (λ, x) is a Perron eigenpair. To prove that λ is the *only* Perron eigenvalue, observe that since A is a positive matrix, so is its transpose A^t. Thus Theorem 1.8 applies to A^t as well, and produces what we might call a "left-Perron[6] eigenpair" (μ, y), where μ is a positive eigenvalue for A^t and y a positive eigenvector for μ.

From the associative property of matrix multiplication:

$$\mu(y^t x) = (y^t A)x = y^t(Ax) = y^t(\lambda x) = \lambda(y^t x).$$

Now $y^t x$, being the dot product of the positive column vectors y and x, is > 0, thus $\mu = \lambda$. This establishes the uniqueness of Perron eigenvalues, since if λ' is another Perron eigenvalue for A then $\lambda' = \mu = \lambda$.

To show that the Perron eigenvalue of A is the *largest* eigenvalue, let $r(A)$ denote the *spectral radius* of A, i.e.,

$$r(A) = \max\{|\gamma| : \gamma \text{ is an eigenvalue of } A\},$$

where on the right-hand side we allow *all* eigenvalues of A, even the complex ones!

Claim. $r(A)$ *is the Perron eigenvalue of A.*

Proof of Claim. We wish to show that if λ is the unique Perron eigenvalue of A and μ is an eigenvalue (real or complex) of A, then $|\mu| \leq \lambda$. To this end, let x be a Perron eigenvector for λ. Suppose μ is an eigenvalue of A and w a corresponding eigenvector, so $w \in \mathbb{C}^N \setminus \{0\}$ and $Aw = \mu w$. Let $|w|$ denote the column vector whose j-th entry is the absolute value of the corresponding entry of w. Then since the entries of A are non-negative:

$$A|w| \geq |Aw| = |\mu w| = |\mu| \, |w|, \tag{1.6}$$

where the inequality is coordinatewise. Now let y be a Perron eigenvector for A^t, so $y > 0$ and, since the Perron eigenvalues for A and A^t coincide, $y^t A = \lambda y^t$. Upon left-multiplying both sides of (1.6) by the positive vector y^t we obtain

$$\lambda y^t |w| = (y^t A)|w| \geq y^t |Aw| = |\mu| y^t |w| \tag{1.7}$$

which implies (since the scalar $y^t |w|$ is > 0) that $\lambda \geq |\mu|$. Thus $\lambda = r(A)$, as desired.

So far we know that the positive matrix A has exactly one Perron eigenvalue, namely the spectral radius $r(A)$. Now we want to show that there is just one Perron eigenvector for this eigenvalue.

Suppose to the contrary that there are two distinct Perron eigenvectors x and y for $r(A)$, so that the pair x, y is linearly independent in $\mathbb{R}^N$.

Claim: There exists a real number $a > 0$ such that the vector $w = ax - y$ is ≥ 0 but not > 0 (i.e., it has at least one coordinate equal to zero).

---

[6] The terminology "left-Perron" comes from the fact that $y^t A = \mu y^t$, i.e., the row vector y^t is a "left eigenvector" for A with "left eigenvalue" μ.

Granting this Claim: By linear independence, $w \neq 0$, and clearly $Aw = r(A)w$. Every entry of the matrix A is positive, and those of w are non-negative and not all zero, so Lemma 1.9 assures us that $r(A)w = Aw > 0$. But $r(A) > 0$ by Theorem 1.8, so $w > 0$: a contradiction.

Proof of Claim. It remains to find the positive constant a. For this, let ξ_j denote the j-th coordinate of the Perron vector x, and η_j the j-th coordinate of y. We're looking for $a \in \mathbb{R}$ such that $a\xi_j \geq \eta_j$ for all j, and $a\xi_k = \eta_k$ for *some* k. Since no coordinate of x is zero, we can rewrite our criteria as: $a \geq \eta_j/\xi_j$ for all j, and $a = \eta_k/\xi_k$ for some k; in other words the positive real number $a = \max_j \eta_j/\xi_j$ does the job. □

Corollary 1.11. *If A is a positive $N \times N$ stochastic matrix, then there is a positive vector $x \in \Pi_N$ such that $Ax = x$. The vector x is, up to scalar multiples, the unique non-zero fixed point of A.*

Proof. Since A is stochastic, Theorem 1.6 supplies a vector $x_0 \in \mathbb{R}^N_+ \backslash \{0\}$ with $Ax_0 = x_0$. Thus $x = x_0/\|x_0\|_1$ is a fixed point of A that lies in Π_N. Since A is also positive, Perron's Theorem (Theorem 1.10) guarantees that x is positive and is (up to scalar multiples) the unique eigenvector of A for the eigenvalue 1. □

> *Exercise* 1.10 (Uniqueness of the Perron Eigenvector). Extend the argument above to show that if A is a positive $N \times N$ matrix and x is a Perron vector for $r(A)$, then the real eigenspace $\{w \in \mathbb{R}^N : Aw = r(A)w\}$ is one dimensional. Then show that the corresponding *complex* eigenspace is also one dimensional.

> *Exercise* 1.11 (Loneliness of the Perron Eigenvalue). Show that if A is an $N \times N$ positive matrix then its Perron eigenvalue is the only eigenvalue on the circle $\{z \in \mathbb{C} : |z| = r(A)\}$.
>
> *Suggestion:* Suppose μ is an eigenvalue of A (real or complex) with $|\mu| = r(A)$. Let $w \in \mathbb{C}^N \backslash \{0\}$ be a μ-eigenvector of A. Without loss of generality we may assume that some coordinate of w is positive. Our assumption that $|\mu| = r(A)$ implies that there is equality in (1.7), and this implies that $\sum_j a_{i,j}|w_j| = |\sum_j a_{i,j}w_j|$ for all indices i. Conclude that $w_j \geq 0$ for all j, hence $\mu > 0$.

1.9 The Google Matrix

We'd like to apply Corollary 1.11 to the problem of ranking internet pages. Unfortunately, the modified hyperlink matrix H_1 we created at the end of Sect. 1.4 (page 8), while stochastic, is far from positive; in fact we noted that "almost all" of its entries are zero. But all is not lost: a simple modification of H_1 shows that a reasonable model of web-surfing can arise from a positive stochastic matrix.

Let E denote the $N \times N$ matrix, each of whose entries is 1. Fix a "damping factor" d with $0 < d < 1$, and let G denote the "Google Matrix"

$$G = dH_1 + \frac{(1-d)}{N}E.$$

Since the matrices H_1 and $\frac{1}{N}E$ are both stochastic, so is G (cf., Exercise 1.9, page 12). Furthermore $G > 0$, so Corollary 1.11 guarantees a fixed point $w > 0$ that is unique up to scalar multiples. Thus G provides a unique ranking of web pages.

Why does G provide a reasonable model for web-surfing? Recall that we've already noted how the modified hyperlink matrix H_1 represents what might be termed a "semi-deterministic" model for web-surfing, wherein the surfer *at a given page* chooses randomly among its outlinks with uniform probability and, if there are no outlinks, chooses randomly, again with uniform probability, from all possible web pages. In this vein, the matrix $\frac{1}{N}E$ represents a purely random surfing strategy, wherein our surfer at a given page ignores all links and moves to another page (or stays put) with probability $1/N$. Thus the matrix G models the behavior of a surfer who, at a given page, chooses the next one using the semi-deterministic method with probability d, and the purely random one with probability $1 - d$. Google's early experiments indicated that $d = 0.85$ could provide a reasonable start on a web-surfing model [17].

The Elephant in the Room. Let's not forget that G is a huge matrix: $N \times N$ with N in the billions! Thanks to Brouwer and Perron we now know that G produces a unique ranking of web pages, but it's still not clear how to effectively compute this ranking. The fixed-point theorem of Chap. 3 will show us a way that is simple—at least in principle—to do this.

Notes

The Banach Mapping Theorem. Theorem 1.1 first appeared (for one-to-one mappings) in Banach's paper [7]. I learned the result from John Erdman, who presented it, along with its application to the Schröder–Bernstein Theorem, in a seminar at Portland State. The same proof has recently been found independently by Ming-Chia Li [69]. We'll encounter this result again in Chap. 11 when we take up the remarkable subject of paradoxical decompositions.

The Knaster–Tarski Theorem. In [116, page 286] Tarski points out that in the 1920s he and Knaster discovered Theorem 1.2, with Knaster publishing the result in [63]. Tarski goes on to say that he found a generalization to "complete lattices" and lectured on it and its applications during the late 1930s and early 1940s before finally publishing his results in [116].

The Brouwer Fixed-Point Theorem. This result (Theorem 1.3) appeared in [18, 1912], where it was proved using topological methods developed by Brouwer. It is one of the most famous and widely applied theorems in mathematics; see [91] for an exhaustive survey of the legacy of this result, and [21, Chap. 1] for a popular exposition.

Positive matrices. The arguments used here to prove Theorem 1.10, the famous theorem of Perron (1907), follow those of [30, Chap. 2]. In 1912 Frobenius extended Perron's results to certain matrices with non-negative entries. The resulting "Perron–Frobenius" theory is the subject of ongoing research, with an enormous literature spanning many scientific areas. For more on this see, e.g., [30, 72] or[76].

Chapter 2
Brouwer in Dimension Two

THE BROUWER FIXED-POINT THEOREM VIA SPERNER'S LEMMA

Overview. In dimension two the Brouwer Fixed-Point Theorem states that every continuous mapping taking a closed disc into itself has a fixed point. In this chapter we'll give a proof of this special case of Brouwer's result, but for triangles rather than discs; closed triangles are homeomorphic to closed discs (Exercise 2.2 below) so our result will be equivalent to Brouwer's. We'll base our proof on an apparently unrelated combinatorial lemma due to Emanuel Sperner, which—in dimension two—concerns a certain method of labeling the vertices of "regular" decompositions of triangles into subtriangles. We'll give two proofs of this special case of Sperner's Lemma, one of which has come to serve as a basis for algorithms designed to approximate Brouwer fixed points.

Prerequisites. Undergraduate real analysis: compactness and continuity in the context of $\mathbb{R}^2$.

2.1 Sperner's Lemma

Throughout this discussion, "triangle" means "closed triangle," i.e., the convex hull of three points in Euclidean space that don't all lie on the same straight line. A "regular decomposition" of a triangle is a collection of subtriangles whose union is the original triangle and for which the intersection of two distinct subtriangles is either a vertex or a complete common edge. Figure 2.1 below illustrates both a regular and an irregular decomposition of a triangle into subtriangles.

A "Sperner Labeling" of the subvertices (the vertices of the subtriangles) in a regular decomposition is an assignment of labels "1," "2," or "3" to each subvertex in such a way that:

(a)   No two vertices of the original triangle get the same label (i.e., all three labels get used for the original vertices),

© Springer International Publishing Switzerland 2016
J.H. Shapiro, *A Fixed-Point Farrago*, Universitext,
DOI 10.1007/978-3-319-27978-7_2

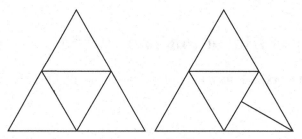

Fig. 2.1 Regular (*left*) and irregular (*right*) decomposition of a triangle into subtriangles

(b)   Each subvertex lying on an edge of the original triangle gets labels drawn only
       from the labels of that edge, e.g., subvertices on the original edge labeled "1"
       and "2" (henceforth: a "{1,2} edge") get only the labels "1" or "2," but with
       no further restriction.
(c)   Subvertices lying in the interior of the original triangle can be labeled without
       restriction.

We'll call a subtriangle whose vertices have labels "1," "2," and "3" a *completely labeled subtriangle*. Figure 2.2 shows a regular decomposition of a triangle
into Sperner-labeled subtriangles, five of which (the shaded ones) are completely
labeled.

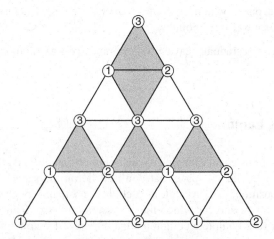

Fig. 2.2 A Sperner-labeled regular decomposition into subtriangles

Theorem 2.1 (Sperner's Lemma for Dimension Two). *Every Sperner-labeled regular decomposition of a triangle has an odd number of completely labeled subtriangles; in particular there is at least one.*

The One dimensional case. Here, instead of triangles split "regularly" into subtriangles, we just have a closed finite line segment split into finitely many closed subsegments which can intersect in at most a common endpoint. One end of the original segment is labeled "1" and the other is labeled "2." The remaining subsegment endpoints get these labels without restriction.

Sperner's Lemma for this situation asserts that: *There is an odd number of subsegments (in particular, at least one!) whose endpoints get different labels.*

To prove this let's imagine moving from the one-labeled endpoint of our initial interval toward the two-labeled one. If there are no subintervals, we're done. Otherwise there has to be a first subinterval endpoint whose label switches from "1" to "2," thus yielding a completely labeled subinterval with final endpoint "2." At the next switch, if there is one, the initial endpoint is "2" and the final endpoint is "1," thus yielding another completely labeled subinterval which must, somewhere further on the line, have an oppositely labeled companion (else we'd never be able to end up with the final subinterval labeled "2"). Thus there must be an odd number of completely labeled subintervals. □

The Two dimensional case. We start with a triangle Δ regularly decomposed into a finite collection of subtriangles $\{\Delta_j\}$. Let $v(\Delta_j)$ denote the number of "$\{1,2\}$-labeled edges" belonging to the boundary of Δ_j, and set $S = \sum_j v(\Delta_j)$. We'll compute S in two different ways:

By counting edges. If a $\{1,2\}$-labeled edge of Δ_j does not belong to the boundary of Δ then it belongs to exactly one other subtriangle. If a $\{1,2\}$-labeled edge of Δ_j lies on the boundary of Δ, then that edge belongs to no other subtriangle. Thus S is twice the number of "non-boundary" $\{1,2\}$-labeled edges plus the number of "boundary" $\{1,2\}$-labeled edges. But by the one dimensional Sperner Lemma, the number of boundary $\{1,2\}$-labeled edges is odd. Thus S *is odd.*

By counting subtriangles. Each completely labeled subtriangle has exactly one $\{1,2\}$-labeled edge. All the others have either zero or two such edges. Thus the odd number S is the number of completely labeled subtriangles plus twice the number of subtriangles with $\{1,2\}$ edges, hence our Sperner-labeled regular decomposition of Δ has an odd number of completely labeled subtriangles.

2.2 Proof of Brouwer's Theorem for a Triangle

We may assume, without loss of generality (see the exercise below), that our triangle Δ is the standard simplex Π_3 of $\mathbb{R}^3$ (see Definition 1.7). Fix a continuous self-map f of Δ; for each $x \in \Delta$ write $f(x) = (f_1(x), f_2(x), f_3(x))$. Thus for each index $j = 1, 2, 3$ we have a continuous "coordinate function" $f_j \colon \Delta \to [0,1]$ with $f_1(x) + f_2(x) + f_3(x) = 1$ for each $x \in \Delta$.

A Sperner labeling induced by f. Consider a regular decomposition of Δ into sub-triangles and suppose f fixes no subvertex (if f fixes a subvertex, we are done). Then f determines a Sperner labeling of subvertices in the following manner. Fix a subtriangle vertex p. Since $f(p) \neq p$, and since both p and $f(p)$ have non-negative coordinates that sum to 1, at least one coordinate of $f(p)$ is strictly less than the corresponding coordinate of p. Choose such a coordinate and use its index to label the subvertex p.

In this way the three original vertices $e_1 = (1,0,0)$, $e_2 = (0,1,0)$, and $e_3 = (0,0,1)$, get the labels "1," "2," and "3," respectively. For example, $f(e_1) \neq e_1$, so the first coordinate of $f(e_1)$ must be strictly less than 1, and similarly for the other two vertices of Δ. Each vertex on the $\{1,2\}$ edge of $\partial\Delta$ (the line segment joining e_1 to e_2) has third coordinate zero, so this coordinate cannot strictly decrease when that vertex is acted upon by f. Thus (since that vertex is not fixed by f) at least one of the other coordinates must strictly decrease, so each vertex on the $\{1,2\}$-edge gets only the labels "1" or "2," as required by Sperner labeling. Similarly for the other edges of $\partial\Delta$; the vertices on the $(2,3)$-edge get only labels "2" and "3," and the vertices on the $(1,3)$-edge get only labels "1" and "3." No further checking is required for the labels induced by f on the interior vertices; Sperner labeling places no special restrictions here. In this way f determines, for each regular subdivision of Δ, a Sperner labeling of the subvertices (note that the continuity assumed for f has not yet been used).

Approximate fixed points for f. Let $\varepsilon > 0$ be given. We're going to show that our continuous self-map f of Δ has an *ε-approximate fixed point*, i.e., a point $p \in \Delta$ such that $\|f(p) - p\|_1 \leq \varepsilon$. Here $\|x\|_1$ is the "one-norm" of $x \in \mathbb{R}^3$, as defined by Eq. (1.5) (page 11). Being continuous on the compact set Δ, the mapping f is *uniformly* continuous there, so there exists $\delta > 0$ such that $x, y \in \Delta$ with $\|x - y\|_1 < \delta$ implies $\|f(x) - f(y)\|_1 < \varepsilon/8$. Upon decreasing δ if necessary we may assume that $\delta < \varepsilon/8$. Now suppose Δ is regularly decomposed into subtriangles of $\|\cdot\|_1$-diameter $< \delta$. If some subvertex of this decomposition is a fixed point of f, we're done. Suppose otherwise. Thus f creates a Sperner labeling of the subvertices of this decomposition. Let Δ_ε be a completely labeled subtriangle, as promised by Sperner's Lemma.

Claim. Δ_ε contains an ε-approximate fixed point.

Proof of Claim. Let p, q, and r be the vertices of Δ_ε, carrying the labels "1," "2," and "3," respectively, so that $f_1(p) < p_1$, $f_2(q) < q_2$, and $f_3(r) < r_3$. Thus:

$$\|p - f(p)\|_1 = \underbrace{p_1 - f_1(p)}_{>0} + |p_2 - f_2(p)| + |p_3 - f_3(p)|$$

$$= p_1 - f_1(p) + |q_2 - f_2(q) + p_2 - q_2 + f_2(q) - f_2(p)|$$

$$+ |r_3 - f_3(r) + p_3 - r_3 + f_3(r) - f_3(p)|$$

$$\leq \underbrace{p_1 - f_1(p)}_{>0} + \underbrace{q_2 - f_2(q_2)}_{>0} + \underbrace{r_3 - f_3(r_3)}_{>0}$$

$$+ |p_2 - q_2| + |f_2(q) - f_2(p)|$$

$$+ |p_3 - r_3| + |f_3(r) - f_3(p)|,$$

so $\|p - f(p)\| \leq A + B$, where

$$A = [p_1 - f_1(p)] + [q_2 - f(q_2)] + [r_3 - f(r_3)]$$

which is > 0 since this is true of each bracketed term, and

$$B = |p_2 - q_2| + |f_2(q) - f_2(p)| + |p_3 - r_3| + |f_3(r) - f_3(p)|. \qquad (2.1)$$

Now each summand on the right-hand side of (2.1) is $< \varepsilon/8$, hence $B < \varepsilon/2$. As for A, the same "adding-zero trick" we used above yields

$$A = \underbrace{p_1 + p_2 + p_3}_{=1} - \underbrace{f_1(p) + f_2(p) + f_3(p)}_{=1}$$

$$+ [q_2 - p_2] + [f_2(p) - f_2(q)]$$

$$+ [r_3 - p_3] + [f_3(p) - f_3(r)].$$

On the right-hand side of this equation, the top line equals zero and each bracketed term has absolute value $< \varepsilon/8$, so by the triangle inequality, $A < \varepsilon/2$. These estimates on A and B yield $\|p - f(p)\|_1 < \varepsilon$, the vertex p of Δ_ε is an ε-approximate fixed point of f. $\square$

The same argument shows that the other two vertices of Δ_ε are also ε-approximate fixed points of f; the triangle inequality shows that *every* point of Δ_ε is a $\frac{5}{4}\varepsilon$-approximate fixed point.

A fixed point for f. So far we know that our self-map f of Δ has an ε-approximate fixed point for every $\varepsilon > 0$. In particular, for each positive integer n there is a $1/n$-approximate fixed point x_n. Since Δ is compact there is a subsequence (x_{n_k}) convergent to some point $x \in \Delta$. By the triangle inequality for the norm $\| \cdot \|_1$:

$$\|x - f(x)\|_1 \leq \|x - x_{n_k}\|_1 + \|x_{n_k} - f(x_{n_k})\|_1 + \|f(x_{n_k}) - f(x)\|_1$$

On the right-hand side of this inequality, as $k \to \infty$:

The first summand $\to 0$ (since $x_{n_k} \to x$).
Therefore the third summand $\to 0$ by the continuity of f.
The second summand $\to 0$ because it's $< 1/n_k$.

Conclusion: $\|x - f(x)\|_1 = 0$, hence $f(x) = x$, as desired. $\square$

The argument above works much more generally to prove:

Lemma 2.2 (The Approximate-Fixed-Point Lemma). *Suppose (X,d) is a compact metric space and $f: X \to X$ is a continuous map. Suppose that for every $\varepsilon > 0$ there exists a point $x_\varepsilon \in X$ with $d(f(x_\varepsilon), x) \leq \varepsilon$. Then f has a fixed point.*

Proof. Exercise: generalize the proof given above for the metric induced by the one-norm to arbitrary metrics.   □

> *Exercise* 2.1. Here's another way to produce fixed points from completely labeled subtriangles. Make a regular decomposition of Δ into subtriangles of diameter $< 1/n$. For this decomposition of Δ, use f to Sperner-label the subvertices, and let Δ_n be a resulting completely labeled subtriangle. Denote the vertices of Δ_n by $p^{(n)}$, $q^{(n)}$, and $r^{(n)}$, using the previous numbering scheme so that $f_1(p^{(n)}) \leq p_1^{(n)}$, etc. Show that it's possible to choose a subsequence of integers $n_k \nearrow \infty$ such that the corresponding subsequences of p's, q's, and r's all converge. Show that these three subsequences all converge to the same point of Δ, and that this point is a fixed point of f.

> *Exercise* 2.2. Show that every triangle is homeomorphic to a closed disc.
>
> *Suggestion:* First argue that without loss of generality we can suppose that our triangle T lies in $\mathbb{R}^2$, contains the origin in its interior, and is contained in the closed disc D of radius 1 centered at the origin. Then each point $z \in T \setminus \{0\}$ is uniquely represented as $z = r\zeta$ for $\zeta \in \partial D$ and $r > 0$. Let $w = \rho\zeta$ be the point at which the line through the origin and z intersects ∂T. Show that the map that fixes the origin and takes $z \neq 0$ to $(r/\rho)\zeta$ is a homeomorphism of T onto D.

2.3 Finding Fixed Points by "Walking Through Rooms"

Finding fixed points "computationally" amounts to finding an algorithm that produces sufficiently accurate approximate fixed points. Thanks to the work just done in Sect. 2.2, an algorithm for finding a completely labeled subtriangles will do the trick. Here's an alternate proof of Sperner's Lemma that speaks to this issue.

Imagine our triangle Δ to be a house, and that the subtriangles of a regular subdivision are its rooms. Given a Sperner labeling of the subvertices that arise from this decomposition, think of each $\{1,2\}$-labeled segment of a subtriangle boundary as a door; these are the only doors. For example, a $\{1,2,2\}$-labeled subtriangle has two doors, some rooms have no doors (e.g., those with no subvertex labeled "2"); the completely labeled subtriangles are those rooms with exactly one door.

Now imagine that you are outside the house. *There is a door to the inside*; the Sperner labeling of the subvertices induces on the original $\{1,2\}$ edge a one dimensional Sperner labeling which, by the $N = 1$ case of Sperner's Lemma, must produce a $\{1,2\}$-labeled subinterval. Go through this door. Once inside either the room you're in has no further door, in which case you're in a completely labeled subtriangle, or there is another door to walk through. Keep walking, subject to the rule that you can't pass through a door more than once (i.e., the doors are "trapdoors"). There are two possibilities. Either your walk terminates in a completely labeled room, in which case you're done, or it doesn't, in which case you find yourself back outside the house. In that case, you've used up two doors on the $\{1,2\}$ edge of Δ: one to go into the house, and the other to come back out. But according to the one dimensional Sperner Lemma, there are an odd number of such boundary doors, so there's one you haven't used. Re-enter the house. Continue. In a finite number of steps you must encounter a room with just one door: a completely labeled one. $\square$

Figure 2.3 below illustrates this process. Starting at point A one travels through three rooms, arriving outside at point B. The process starts again at B, this time terminating at C, inside a completely labeled subtriangle.

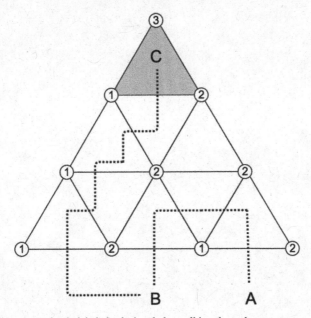

Fig. 2.3 Finding a completely labeled subtriangle by walking through rooms

Notes

Sperner's Lemma, higher dimensions. This result for all finite dimensions appears in Sperner's 1928 doctoral dissertation [111]. In dimensions > 2 the analogue of a triangle is an "N-simplex" in $\mathbb{R}^N$; the convex hull of $N+1$ points of $\mathbb{R}^N$ in "general position," i.e., no point belongs to the convex hull of the others. The analogue of our regular decomposition of a triangle is a "triangulation" of an N-simplex into "elementary sub-simplices," each of which is itself an N-simplex.

Nice descriptions of this generalization occur in [40, Chap. 3, Sect. 4], and in E.F. Su's expository article [113], which also provides a proof of the general Brouwer theorem based on "walking through rooms." Su's article also contains interesting applications of Sperner's Lemma to problems of "fair division."

Walking through rooms. In [113] Su attributes this argument to Kuhn [64] and Cohen [26]. According to Scarf [106], however, the argument has its origin in Lemke's 1965 paper [68]. This technique has been greatly refined to produce useful algorithms for finding approximate fixed points, especially by Scarf, whose survey [106], in addition to providing a nice introduction to the legacy of Sperner and Lemke in the algorithmic search for fixed points, also introduces the reader to the way in which economists view Brouwer's theorem.

Chapter 3
Contraction Mappings

Overview. In this chapter we'll study the best-known of all fixed-point theorems: the Banach Contraction-Mapping Principle, which we'll apply to Newton's Method, initial-value problems, and stochastic matrices.

Prerequisites. Undergraduate-level real analysis and linear algebra. The basics of metric spaces: continuity and completeness.

3.1 Contraction Mappings

The theorem we're going to apply to Newton's Method, Initial-Value Problems, and the Internet was proved by the Polish mathematician Stefan Banach as part of his 1922 doctoral dissertation. Although the setting of Banach's theorem is far more general than that of Brouwer's, the restricted nature of the mappings involved makes its proof a lot simpler.

Banach's theorem is set in a *metric space*: a pair (S,d) where S is a set and d is a "metric" on S, i.e., a function $d: S \times S \to \mathbb{R}_+$ such that for all $x,y,z \in S$

(m1) $d(x,y) = 0$ iff $x = y$,
(m2) $d(x,y) = d(y,x)$, and
(m3) $d(x,z) \leq d(x,y) + d(y,z)$.

The last property is called, for obvious reasons, "the triangle inequality."

Example. Let S be $\mathbb{R}^N$, or a subset thereof, and take $d(x,y)$ to be the Euclidean distance between x and y: $d(x,y) = \|x - y\|$. Alternatively d could be the distance on $\mathbb{R}^N$ induced in the same way by the *one-norm* introduced in the proof of Theorem 1.6. As we pointed out there, the two metrics are equivalent in that they have the same convergent sequences.

© Springer International Publishing Switzerland 2016
J.H. Shapiro, *A Fixed-Point Farrago*, Universitext,
DOI 10.1007/978-3-319-27978-7_3

The mappings addressed by Banach's Principle are called *strict contractions*.[1] To say $F\colon S \to S$ is one of these means that there is a positive "contraction constant" $c < 1$ for which

$$d(F(x), F(y)) \le cd(x,y) \qquad \forall\, x, y \in S. \tag{3.1}$$

Clearly every strict contraction is continuous on S.

Definition 3.1. A *Cauchy sequence* in a space with metric d is a sequence (x_n) such that: For each $\varepsilon > 0$ there is a positive integer $N = N(\varepsilon)$ such that $d(x_n, x_m) < \varepsilon$ whenever the indices m and n are larger than N. A *complete metric space* is one in which every Cauchy sequence converges.

Theorem 3.2 (The Banach Contraction-Mapping Principle). *Suppose (S,d) is a complete metric space and $F\colon S \to S$ is a strict contraction. Then F has a* unique *fixed point, and every iterate sequence converges to this point.*

We'll prove this shortly; first, a few comments.

Iterate sequence. Recall that, for a mapping F taking a set S into itself, the *iterate sequence* starting at $x_0 \in S$ is (x_n) where $x_{n+1} = F(x_n)$ for $n = 0, 1, 2, \ldots$.

Uniqueness. If (S,d) is a metric space on which F is a strict contraction and $p \in S$ is a fixed point of F, then *there can be no other fixed point.*

Proof. If $q \in S$ is also a fixed point of f then

$$d(p,q) = d(F(p), F(q)) \le cd(p,q).$$

Since $0 < c < 1$ we must have $d(p,q) = 0$, whereupon condition (m1) in the definition of "metric" guarantees that $p = q$. $\square$

"Non-strict" contractions. If in Theorem (3.1) we merely assume that the contraction constant c is 1, then:

–   *Existence* can fail. Example: $F(x) = x + 1$ defined on the real line.
–   *Uniqueness* can also fail. Example: the identity map on a metric space with more than one point.

> *Exercise* 3.1 (Necessity of completeness). Give an example of an *incomplete* metric space on which there is a strict contraction with no fixed point.

Fixed points and iterate sequences. We contended in Sect. 1.2 (page 4) that if the iteration of Newton's method for an appropriate function f were to converge, then that limit had to be a root of f (i.e., a fixed point of the Newton function of f). The next result justifies this contention in a far more general setting.

Proposition 3.3. *If (S,d) is a metric space, $F\colon S \to S$ is continuous, and x_0 is a point of S for which the iterate sequence $\{x_0, F(x_0), F(F(x_0)), \ldots\}$ converges, then the limit of that sequence has to be a fixed point of F.*

---

[1] These are often just called "contractions"; the terminology here is more in keeping with conventions used in (linear) operator theory.

Proof. Suppose the iterate sequence (x_n) of x_0 converges to $p \in S$, i.e., $\lim_n d(x_n, p) = 0$. Then the continuity of F insures that $x_{n+1} = F(x_n) \to F(p)$. Also $\lim_n d(x_{n+1}, p) = 0$, i.e., $x_{n+1} \to p$, so (because limits in metric spaces are unique) $p = F(p)$. $\quad\square$

If we assume further that F is a strict contraction, then there results a very strong converse.

Proposition 3.4. *Suppose F is a strict contraction on a metric space. If p is a fixed point of F then* every *iterate sequence converges to p.*

Proof. Let c denote the contraction constant of the mapping F, so $0 < c < 1$ and F satisfies (3.1) above. Fix $x_0 \in S$ and define the iterate sequence (x_n) in the usual way: $x_1 = F(x_0), \dots, x_n = F(x_{n-1}), \dots$. Then

$$d(x_n, p) = d(F(x_{n-1}), F(p)) \le cd(x_{n-1}, p) \le \dots \le c^n d(x_0, p),$$

so $d(x_n, p) \to 0$ as $n \to \infty$, i.e., (x_n) converges to p. $\quad\square$

> *Exercise* 3.2 (Lessons from a simple initial-value problem). For the initial-value problem
> (IVP) $y' = y$, $y(0) = 1$, write down the integral operator T on $C(\mathbb{R})$ defined on page 4 by
> Eq. (IE), and compute explicitly the iterate sequence that has $y_0 \equiv 1$ as its initial function.
> On which intervals $[-a, a]$ does this iterate sequence converge uniformly to a solution of the
> IVP? For which of these intervals does the Contraction-Mapping Principle guarantee such
> convergence?

Proof of the Contraction-Mapping Principle. In view of Proposition 3.4 only one strategy will work: fix a point $x_0 \in S$ and prove that its iterate sequence (x_n) converges. By Proposition 3.3 this limit must be a fixed point.

Since our metric space is complete it's enough to show that (x_n) is a Cauchy sequence. To this end, consider a pair of indices $m < n$ and use the triangle inequality to observe that

$$d(x_n, x_m) \le \sum_{j=m}^{n-1} d(x_{j+1}, x_j).$$

From the strict contractiveness of F:

$$d(x_{j+1}, x_j) = d(F(x_j), F(x_{j-1}) \le cd(x_j, x_{j-1}) \le \dots \le c^j d(x_1, x_0),$$

whereupon (since $c < 1$)

$$d(x_m, x_n) \le \sum_{j=m}^{n-1} c^j d(x_1, x_0) = d(x_1, x_0) \sum_{j=m}^{\infty} c^j = \frac{d(x_1, x_0)}{1 - c} c^m.$$

Now given $\varepsilon > 0$, we may choose N so that $\frac{d(x_1, x_0)}{1-c} c^N < \varepsilon$, which insures, by the above chain of inequalities, that $N \le m < n \Rightarrow d(x_m, x_n) < \varepsilon$, hence our iterate sequence (x_n) is indeed Cauchy. $\quad\square$

The Contraction-Mapping Principle seems to be a perfect theorem: easy to prove and widely applicable. However there is a catch: proving a given mapping to be a strict contraction usually requires some work—as you'll see in the next few sections.

3.2 Application: Stochastic Matrices/Google

In Sect. 1.9 we introduced the "Google matrix" G, a stochastic matrix with entries all positive, and observed with the help of the Brouwer Fixed-Point Theorem (a key step in our proof of Perron's Theorem) that G has an essentially unique positive fixed point whose coordinates rank internet web pages.

There remains, however, the problem of proposing an algorithm for actually *finding* this fixed point. Recall that the application of Brouwer/Perron to the Google matrix ultimately rested on the stochasticity of that matrix, which implied that G (indeed each $N \times N$ stochastic matrix) maps the standard N-simplex Π_N continuously into itself. The positivity of G then guaranteed the uniqueness of its fixed point.

The generalization to $\mathbb{R}^N$ of the "walking-through-rooms" proof of Brouwer's theorem set out for $N = 2$ in Sect. 2.3 could provide the basis for an algorithm that approximates the desired fixed point. On the other hand, Banach's theorem has built into it a scheme that—at least in theory—is easily implemented: *Use iterate sequences to approximate fixed points.* However to be certain that this will work we need each positive stochastic matrix to induce a *strict* contraction on its standard simplex (so far we've established only "non-strict" contractivity: Exercise 1.5).

Does stochasticity imply strict contractivity? Is this too much to ask? Read on!

Theorem 3.5. *Every $N \times N$, positive, stochastic matrix induces a strict contraction on the standard simplex Π_N, taken in the metric induced by the one-norm.*

Proof. Suppose A is a positive, stochastic, $N \times N$ matrix. We already know (proof of Theorem 1.6) that A takes Π_N into itself. We're claiming that there exists a positive number c strictly less than 1 such that

$$\|Ax - Ay\|_1 \leq c\|x - y\|_1 \qquad (x, y \in \Pi_N). \tag{3.2}$$

Let $a_{i,j}$ denote the matrix A's entry in the i-th row and j-th column. Since each of these numbers is positive we may choose a positive number ε that is strictly less than all of them. Since each column of A sums to 1 we know that $N\varepsilon < 1$ (Proof: for j an index, $1 = \sum_i a_{i,j} > N\varepsilon$). Thus we may form the new $N \times N$ matrix B, whose (i, j)-entry is

$$b_{i,j} = \frac{a_{i,j} - \varepsilon}{1 - N\varepsilon}.$$

Clearly B is a positive matrix, and it's easy to check that B is stochastic. Now

$$A = (1 - N\varepsilon)B + \varepsilon E$$

where E is the $N \times N$ matrix, all of whose entries are 1.

Claim. A satisfies (3.2) with $c = (1 - N\varepsilon)$.

Proof of Claim. Since $N\varepsilon$ lies strictly between 0 and 1, so does c. What makes this argument work is the fact that if $x \in \Pi_N$ then Ex is the vector in $\mathbb{R}^N$, each of whose coordinates is the sum of the coordinates of x, namely 1. In particular if x and y belong to Π_N then $Ex = Ey$, whereupon

$$Ax - Ay = c(Bx - By) + \varepsilon(Ex - Ey) = c(Bx - By).$$

By Exercise 1.5, every $N \times N$ stochastic matrix induces, in the 1-norm, a (possibly non-strict) contraction on $\mathbb{R}^N$, so from the last equation and the linearity of B:

$$\|Ax - Ay\|_1 = \|c(Bx - By)\|_1 = c\|B(x-y)\|_1 \le c\|x - y\|_1 \quad (x, y \in \Pi_N),$$

which proves the Claim, and with it the theorem. $\qquad\square$

Corollary 3.6. *If A is an $N \times N$ positive stochastic matrix, then its (unique) Perron eigenvector is the limit of the iterate sequence of each initial point $x_0 \in \Pi_N$.*

In particular, the unique ranking of web pages produced by the Google matrix can be computed by iteration. For $x_0 \in \Pi_N$ the iterate sequence of Corollary 3.6 is (x_n), where

$$x_n = Ax_{n-1} = A^2 x_{n-2} = \ldots = A^n x_0 \qquad (n = 1, 2, \ldots).$$

For this reason the approximation scheme of the Corollary is called *power iteration*; it is used widely in numerical linear algebra for eigen-value/vector approximation.

3.3 Application: Newton's Method

Suppose f is a real-valued function defined on a finite, closed interval $[a, b]$ of the real line, and that we know f has a root somewhere in the open interval (a, b). We're going to use the Contraction-Mapping Principle to show that, under suitable hypotheses on f, Newton's method for each appropriate starting point converges to this root.

More precisely, suppose $f \in C^2(I)$ with f' never zero on I, and suppose f has different signs at the endpoints of I; say (without loss of generality) $f(a) < 0$ and $f(b) > 0$. Then f has a unique root x^* in the interior (a, b) of I. Under these hypotheses we have

Theorem 3.7. *There exists $\delta > 0$ such that for every x_0 in $[x^* - \delta, x^* + \delta]$, Newton's method with starting point x_0 converges to x^*.*

In other words, under reasonable hypotheses on f: for starting points close enough to a root of f the iterate sequence for the Newton function

$$F(x) = x - \frac{f(x)}{f'(x)} \qquad (x \in I)$$

will converge to that root.

Proof. Let M denote the maximum of $|f''(x)|$ as x ranges through I, and let m denote the corresponding minimum of $|f'(x)|$. By the continuity of f'', and the hypothesis that f' never vanishes on I, we know that M is finite and $m > 0$.

Differentiation of F via the quotient rule yields

$$F'(x) = \frac{f(x)\,f''(x)}{f'(x)^2} \qquad (x \in I)$$

which, along with our bounds on f' and f'', provides the estimate

$$|F'(x)| \le \frac{M}{m^2}|f(x)| \qquad (x \in I_\delta).$$

Thus, upon shrinking δ enough to insure that

$$|f(x)| \le \frac{m^2}{2M} \qquad \text{for} \qquad |x - x^*| < \delta$$

(possible because f is continuous at x^* and takes the value zero there) we see that $|F'(x)| \le 1/2$ for each $x \in I_\delta = [x^* - \delta, x^* + \delta]$. This estimate on F' does the trick! For starters, if $x, y \in I_\delta$ then, along with the Mean-Value Theorem of differential calculus, it shows that

$$|F(x) - F(y)| = |F'(\bar{x})(x - y)| \le \frac{1}{2}|x - y| \qquad \forall x, y \in I_\delta$$

where on the right-hand side of the equality, $\bar{x}$ lies between x and y. Thus F is a strict contraction on I_δ—once we know F maps that interval into itself. But it does, since the same inequality shows that for each $x \in I_\delta$ (upon recalling that the root x^* of f is a fixed point of F):

$$|F(x) - x^*| = |F(x) - F(x^*)| \le \frac{1}{2}|x - x^*| \le \frac{1}{2}\delta < \delta$$

so $F(x) \in I_\delta$, as desired.

Thus Banach's Contraction-Mapping Principle applies to the strict contraction F acting on the complete metric space $I_\delta = [x^* - \delta, x^* + \delta]$, and guarantees that for every starting point in I_δ the corresponding F-iteration sequence converges to the fixed point of F, which must necessarily be the unique root of f in I_δ. $\square$

In the course of this proof we had to overcome a problem that occurs frequently when one seeks to apply Banach's Principle:

> The metric space for which the problem is originally defined is often *not* the one to which you apply Banach's Principle!

For example, the hypotheses of Theorem 3.7 refer to the Newton function F defined on the compact interval (i.e., the complete metric space) I, but the theorem's *proof* depended on cutting this space down to the smaller one I_δ on which F acted as a strict contraction.

We'll see this scenario play out again in the next section, where we'll have to shrink an entire metric space of continuous functions!

3.4 Application: Initial-Value Problems

It's time for a careful treatment of the initial-value problem (IVP) of Sect. 1.3. Recall its form: There is a differential equation plus initial condition

$$y' = f(x,y), \quad y(x_0) = y_0 \qquad \text{(IVP)}$$

with $(x_0, y_0) \in \mathbb{R}^2$ and f assumed initially to be continuous on all of $\mathbb{R}^2$. Here we'll just assume that f is continuous on a closed rectangle $R = I \times H$, where I and H are compact intervals of the real line, I having radius r and center x_0 and H having radius h and center y_0. Thus R is a compact "r by h" rectangle in the plane, centered at the point (x_0, y_0).

We'll operate in the metric space $C(I)$ consisting of real-valued functions that are continuous on I. In the course of our work we'll need to shrink the radius r of I. To keep the notation simple we'll re-assign the original symbols I, r, and R to the newly shrunken objects, taking care to be sure that what we've accomplished in one setting transfers intact to the new one.

Since continuous functions are bounded on compact sets and attain their maxima thereon, we can define on $C(I)$ the "max-norm"

$$\|u\| = \max_{x \in I} |u(x)| \qquad (f \in C(I))$$

and use this to define a metric d by:

$$d(u,v) = \|u-v\| \qquad (u,v \in C(I)).$$

In this metric a sequence converges (resp. is Cauchy) if and only if it converges (resp. is Cauchy) *uniformly* on I. A fundamental property of uniform convergence is that every sequence of functions in $C(I)$ that is uniformly Cauchy on I converges uniformly on I to a function in $C(I)$.[2] Thus the metric space $(C(I), d)$ is *complete*. As in our treatment of Newton's Method, we'll have to find an appropriate subset of $C(I)$ in which to apply Banach's Theorem. We'll break this quest into several steps.

STEP I. $C(I)$ *is too large.* For (IVP) to make sense for a prospective solution $y = u(x)$ we have to make sure that for every $x \in I$ the point $(x, u(x))$ lies in the domain of the function f on the right-hand side of the differential equation in (IVP). We must therefore restrict attention to functions $u \in C(I)$ having graph $y = u(x)$ contained in R, i.e., for which $|u(x) - y_0| \le h$ for every $x \in I$. In metric-space language this means that in order for (IVP) to make sense, our prospective solutions must lie in

---

[2] See [101, Theorems 7.14 and 7.15, pp. 150–151], for example.

$$\overline{B} = \overline{B}(y_0, h) = \{u \in C(I): \|u - y_0\| \le h\},$$

the closed ball in $C(I)$ of radius h, centered at the constant function y_0.

STEP II. *The integral equation.* As we observed in Sect. 1.3, a real-valued function y defined on the interval I satisfies IVP if and only if it satisfies the integral equation

$$y(x) = y_0 + \int_{t=x_0}^{x} f(t, y(t)) \, dt \quad . \quad (x \in I). \tag{IE}$$

The right-hand side of this equation makes sense for every $u \in \overline{B}$, and defines an integral transformation T on $C(I)$ by

$$(Tu)(x) = y_0 + \int_{t=x_0}^{x} f(t, u(t)) \, dt \qquad (u \in \overline{B}, x \in \mathbb{R}). \tag{IT}$$

By an argument entirely similar to the one used in Sect. 1.3 (pages 4 and 5) to prove that (IVP) is equivalent to the problem of finding a fixed point for (IT), we have

Lemma 3.8. *If $u \in \overline{B}$ then Tu is differentiable on I and $(Tu)'(x) = f(x, u(x))$ for every $x \in I$.*

In particular, T maps $\overline{B}$ into $C(I)$.

STEP III. *Insuring that $T(\overline{B}) \subset \overline{B}$.* To use the Banach Contraction-Mapping Principle we must at the very least insure that T maps $\overline{B}$ into itself. For the moment, let's continue to assume only that f is continuous on the rectangle R, and set

$$M = \max\{|f(x, y)|: (x, y) \in R\}.$$

Fix this value of M for the rest of the proof. Although we'll allow ourselves to shrink the horizontal dimension of the rectangle R, we won't be changing the value of M.

Lemma 3.9. *For M as above: if we redefine the interval I to have radius $r \le h/M$ then $T(\overline{B}) \subset \overline{B}$.*

Proof. For $|x - x_0| \le h/M$ we have for each $u \in \overline{B}$:

$$|Tu(x) - y_0| = \left| \int_{t=x_0}^{x} f(t, u(t)) \, dt \right| \le M|x - x_0| \le Mh/M = h.$$

Thus redefining I to have radius $\le h/M$ insures that $\|Tu - y_0\| \le h$ for each $u \in \overline{B}$, i.e., that T maps $\overline{B}$ into itself.   □

STEP IV. *Strict contractivity.* So far we've found how to shrink the original interval I so that the closed ball $\overline{B}$ of radius h in $C(I)$ is mapped into itself by the integral operator T. This ball, being a closed subset of the complete metric space $C(I)$, is itself complete in the metric inherited from $C(I)$. However to apply Banach's Principle we need to know that T is a *strict contraction* on $\overline{B}$. For this we'll assume that the function f, in addition to being continuous on the rectangle R, is also differentiable there with respect to its second variable, and that this partial derivative (call if f_2) is continuous on R.

Our goal now is to show that T is a strict contraction mapping on $\overline{B}$ for some positive $r \leq h/M$. Then Banach's Contraction-Mapping Principle will guarantee a fixed point for T in $\overline{B}$, hence a unique solution therein to the integral equation (IE), and therefore to the initial-value problem (IVP) on the interval $I = [x_0 - r, x_0 + r]$. Once done we'll have proved

Theorem 3.10 (The Picard–Lindelöf Theorem). *Suppose* $(x_0, y_0) \in \mathbb{R}^2$, *U is an open subset of* $\mathbb{R}^2$ *that contains* (x_0, y_0), *and f is a real-valued function that is continuous on U and has thereon a continuous partial derivative with respect to the second variable. Then the initial-value problem* (IVP) *has a unique solution on some nontrivial interval centered at* x_0.

Proof. By the work above we may choose a compact rectangle $R = I \times H$ in U, centered at (x_0, y_0), such that $T(\overline{B}) \subset \overline{B}$ whenever the length of I is sufficiently small. It remains to see how much further we must shrink I in order to achieve strict contractivity for T on $\overline{B}$. To this end let $M' := \max\{|f_2(x,y)| : (x,y) \in R\}$, where the compactness of R and the continuity of f_2 on R guarantee that the maximum exists. Note first that if y_1 and y_2 belong to the interval H with $y_1 \leq y_2$ then the Mean-Value Theorem of differential calculus guarantees for each $x \in I$ that

$$|f(x,y_2) - f(x,y_1)| = |f_2(x,\eta)(y_2 - y_1)| \leq M'|y_2 - y_1| \qquad (3.3)$$

where on the right-hand side of the equality, η lies between y_1 and y_2. Thus if u and v are functions in $\overline{B}$ and $x \in I$, we have upon letting $J(x)$ denote the closed interval between x and x_0:

$$|Tu(x) - Tv(x)| = \left| \int_{J(x)} [f(t, u(t)) - f(t, v(t))] \, dt \right| \leq \int_{J(x)} |f(t, u(t)) - f(t, v(t))| \, dt$$

$$\leq M' \int_{J(x)} |u(t) - v(t)| \, dt \leq M' \|u - v\| \cdot \text{length of } J(x)$$

$$= M' \|u - v\| \cdot |x - x_0| \leq M' r \|u - v\|$$

where the second inequality follows from estimate (3.3). Thus

$$\|Tu - Tv\| \leq M' r \|u - v\| \qquad (u, v \in \overline{B}),$$

so we can insure that T is a strictly contractive self-map of $\overline{B}$ simply by demanding that, in addition to the restriction $r \leq h/M$ already placed on the radius of I, we insure that r be $< 1/M'$. □

Note that the proof given above will still work if the differentiability of f in the second variable is replaced by a "Lipschitz condition"

$$|f(x,y_1) - f(x,y_2)| \leq M'|y_2 - y_1| \qquad ((x,y_1),(x,y_2) \in R).$$

For initial-value problems, the interval of existence/uniqueness promised us by Banach's Principle could be very small (see Exercise 3.2 for an example of this). There is, however, always a *maximal* such interval, and this interval has the property that *the solution's graph over this interval continues out to the boundary of the region on which the function f is defined and satisfies the Picard–Lindelöf hypotheses*. For details see, e.g., [93, Sect. 2.4].

As an illustration of this phenomenon, consider the simple initial-value problem $y' = a(1 + y^2), y(0) = 0$, where $a > 0$. One checks easily that $y = \tan(ax)$ is a solution for which the maximal interval of existence is $(-\frac{\pi}{2a}, \frac{\pi}{2a})$, and a separation-of-variables argument shows that this is the *only* solution. Thus, even though the right-hand side $f(x,y) = a(1 + y^2)$ of this IVP's differential equation is infinitely differentiable (even real-analytic) *on the entire plane,* the solution exists only on a finite interval, which for large a is very small.

Conclusion: In nonlinear situations, singularities can arise "unexpectedly."

Notes

Banach's doctoral dissertation. This is [5]; the Contraction-Mapping Principle is Theorem 6 on page 160 of that paper.

Stochastic matrices. The proof that every positive stochastic matrix induces a strict contraction on its standard simplex (Theorem 3.5) is from [66], where the result is attributed to Krylov and Bogoliubov. The same proof is in [20, Sect. 4, pp. 578–9].

We mentioned that the "power iteration" method of Corollary 3.6 works in more generality. For more on this, see, e.g., [117, Lecture 27].

For the Google matrix G, revisited in Sect. 3.2, there is still the issue of its enormous size. A preliminary discussion of how to handle this can be found in [17].

Initial-value problems. The Picard–Lindelöf Theorem originates in Lindelöf's 1894 paper [70], in which he generalizes earlier work of Picard. In our special case the iteration associated with Banach's Principle is often called "Picard Iteration."

Higher orders, higher dimensions. The restriction of our discussion of initial-value problems to first order differential equations is not as severe as it seems. Consider, for example, the second order problem for an open interval I containing the point x_0:

$$y'' = f(x, y, y'), \quad y(x_0) = y_0, \quad y'(x_0) = y_1 \quad (x \in I).$$

This problem can be rewritten as: $Y' = F(x, Y)$, $Y(x_0) = Y_0$ for $x \in I$, where $Y = (y, y')$ is a function taking I into $\mathbb{R}^2$, $Y_0 = (y_0, y_1)$ is a vector in $\mathbb{R}^2$ (now thought of as a space of row vectors), and $F(x, Y) = (y', f(x, y, y'))$ maps the original domain of f (a subset of $\mathbb{R}^3$) into $\mathbb{R}^2$.

It's not difficult to check that the proof given above for our original "scalar-valued" IVP works almost *verbatim* in the new setting, with the absolute-value norm on the real line replaced in the higher dimensions by the Euclidean one, thus producing a unique solution for the second order IVP. Of course the idea generalizes readily to initial-value problems of order larger than 2.

Newton's Method again. In a similar vein, our analysis of Newton's Method can be generalized to higher dimensions. Suppose the function f maps some open subset G of $\mathbb{R}^N$ into itself, and that $f(p) = 0$ for some point $p \in G$. If we assume that all first and second order partial derivatives of the components of f are continuous, and that the derivative f', which is now a linear transformation on $\mathbb{R}^N$, is nonsingular at every point of G, then, just as in the single-variable case, we can form the "Newton function" $F(x) = x - f'(x)^{-1} f(x)$, where on the right-hand side we see the inverse of the linear transformation $f'(x)$ acting on the vector $f(x)$. A bit more work than before shows that, when restricted to a suitable closed rectangle centered at p, the function F is a strict contraction, so for every point in that rectangle the Newton iteration converges to p.

Part II
From Brouwer to Nash

The next three chapters focus on the Brouwer Fixed-Point Theorem, beginning with an analysis-based argument that proves the theorem in all finite dimensions. Then we'll use Brouwer's theorem to prove John Nash's Nobel Prize winning result on the existence of "Nash Equilibrium" in game theory. Finally we'll prove Kakutani's set-valued extension of Brouwer's theorem, which will lead us to Nash's celebrated "one-page proof" of his theorem. Throughout, the setting will be finite dimensional, the only background needed being a reasonable exposure to undergraduate-level analysis. Ideas from game theory will be carefully motivated.

Chapter 4
Brouwer in Higher Dimensions

The Brouwer Fixed-Point Theorem in All Finite Dimensions

Overview. Having discussed the Brouwer Fixed-Point Theorem (Chap. 1) and proved it for triangles (Chap. 2), we're ready to prove it in every dimension for closed balls and even for compact, convex sets. Our proof will be quite different from that of Chap. 2, with the combinatorics of Sperner's Lemma replaced by methods of analysis.

Prerequisites. Undergraduate-level real analysis, especially calculus of functions of several variables. Some metric-space theory may be helpful, but is not required; all the action takes place in $\mathbb{R}^N$.

4.1 Fixed Points and Retractions

To say that a metric space (S, d) has the "fixed-point property" means that every continuous mapping of the space into itself has a fixed point. Thus the Brouwer Fixed-Point Theorem can be restated:

Theorem 4.1. *For every positive integer N, the closed unit ball of $\mathbb{R}^N$ has the fixed-point property.*

Our proof of Brouwer's theorem will involve reduction to an equivalent result about an important class of mappings called *retractions*. Suppose S is a metric space and A is a subset of S. To say that a continuous mapping $P \colon S \to A$ is a *retraction* of S onto A means that $P(S) = A$ and the restriction of P to A is the identity map on A. When this happens we'll call A a "retract" of S.

Exercise 4.1. A continuous mapping P is a retraction onto its image if and only if $P \circ P = P$.

Perhaps the most familiar example of a retraction is a linear projection taking $\mathbb{R}^N$ onto a subspace. Here are two such examples, where $S = \mathbb{R}^2$, A is the horizontal axis, and $x = (\xi_1, \xi_2)$ is a typical vector in $\mathbb{R}^2$.

© Springer International Publishing Switzerland 2016
J.H. Shapiro, *A Fixed-Point Farrago*, Universitext,
DOI 10.1007/978-3-319-27978-7_4

(a) Let $P(x) = (\xi_1, 0)$. Here P is the *orthogonal* projection of $\mathbb{R}^2$ onto the horizontal axis.

(b) Let $P(x) = (\xi_1 + \xi_2, 0)$. Now P is a $45°$ projection onto the horizontal axis.

Here's an example more relevant to our immediate purposes. Consider a closed annulus in $\mathbb{R}^2$ centered at the origin, having outer radius 1 and some positive inner radius. For x in this annulus let $P(x) = x/|x|$, where $|\cdot|$ denotes the Euclidean norm on $\mathbb{R}^2$. Then P is a continuous map taking the annulus onto its outer boundary, the unit circle, upon which its restriction is the identity map. Thus the unit circle is a retract of the annulus. This example is of interest to us because no such mapping exists for the unit disc:

The unit circle *not* a retract of the closed unit disc.

This follows immediately from the $N = 2$ version of the Brouwer Fixed-Point Theorem (Chap. 2). Indeed, if there *were* a retraction P taking the closed unit disc onto the unit circle, then $Q = -P$ would be a continuous mapping of the disc into itself (more precisely: onto the unit circle), that has no fixed point.

This argument for the disc works just as well for the closed unit ball of $\mathbb{R}^N$ so: *The Brouwer Fixed-Point Theorem for dimension N will show that no closed ball in $\mathbb{R}^N$ can be retracted onto its boundary.* It is the *converse* of this result that will concern us for the rest of this chapter. Our strategy will be to prove, independent of Brouwer's Theorem:

Theorem 4.2 (The No-Retraction Theorem). *For each positive integer N: There is no retraction taking the closed unit ball of $\mathbb{R}^N$ onto its boundary.*

We'll show in the next section that the No-Retraction Theorem implies the Brouwer Fixed-Point Theorem, after which we'll give our "Brouwer-independent" proof of the No-Retraction Theorem.

4.2 "No-Retraction" $\Rightarrow$ "Brouwer"

We've already noted (Chap. 1, Sect. 1.6) that for $N = 1$ Brouwer's Theorem follows from the Intermediate Value Theorem of elementary calculus, so now we'll work in $\mathbb{R}^N$ with $N > 1$, employing the notation $|\cdot|$ for the Euclidean norm in that space.

Suppose the closed unit ball B of $\mathbb{R}^N$ *does not* have the fixed-point property, i.e., that there is a continuous map $f: B \to B$ that has no fixed point. We'll show that f can be used to construct a retraction of B onto its boundary, thus establishing (the contrapositive equivalent of) the result we want to prove.

To visualize this retraction, note that since we're assuming f fixes no point of B we can draw, for each x in B, the directed half-line that starts at $f(x)$ and passes through x. Let $P(x)$ be the point at which this line intersects ∂B, noting that $P(x) = x$ if $x \in \partial B$. Thus P will be the retraction we seek—once we prove its continuity. It seems intuitively clear from Fig. 4.1 that P should be continuous. To prove this without recourse to pictures we need to represent P analytically:

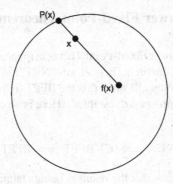

Fig. 4.1 The retraction $P : B \to \partial B$

$$P(x) = x + \lambda(x)u(x) \qquad (x \in B) \qquad (4.1)$$

where $u(x)$ is the unit vector in the direction from $f(x)$ to x:

$$u(x) = \frac{x - f(x)}{|x - f(x)|} \qquad (x \in B) \qquad (4.2)$$

and $\lambda(x)$ is the scalar ≥ 0 chosen to make $|P(x)| = 1$ (so $\lambda(x) = 0$ if $x \in \partial B$).

Since $x - f(x)$ is continuous on B and never zero there, u inherits the continuity of f. As for $\lambda = \lambda(x)$, it is the non-negative solution to the equation

$$0 = |P(x)|^2 - 1 = |x + \lambda u(x)|^2 - 1 = \lambda^2 + 2b\lambda - c \qquad (4.3)$$

where $c = 1 - |x|^2$ and $b = \langle x, u(x) \rangle$, the dot product of the vectors x and $u(x)$. The quadratic equation (4.3) yields solutions $-b \pm \sqrt{b^2 + c}$; since $c \geq 0$ we know that these solutions are real. Since $\sqrt{b^2 + c} \geq \sqrt{b^2} = |b|$ we know that the non-negative solution is the one with the plus sign. Thus

$$\lambda(x) = -\langle x, u(x) \rangle + \sqrt{\langle x, u(x) \rangle^2 + (1 - |x|^2)} \qquad (x \in B), \qquad (4.4)$$

which establishes the desired continuity of λ, and therefore of P. $\qquad\qquad \square$

Exercise 4.2 (More on the "fictitious" unit vector[1] $u(x)$). In the argument above we defined $\lambda(x)$ to take the value zero for $x \in \partial B$, a fact reflected in Eq. (4.3). Note that, thanks to Eq. (4.4) this implies $\langle x, u(x) \rangle \geq 0$ whenever $|x| = 1$. Prove that for all $x \in \partial B$ and $y \in B$ we must have $\langle x, x - y \rangle \geq 0$, with equality if and only if $y = x$. Conclude that in the argument above, $\langle x, u(x) \rangle > 0$ whenever $x \in \partial B$, hence the quantity under the radical sign on the right-hand side of (4.4) is > 0 for every point of B.

---

[1] ..."fictitious" because its existence stems from our assumption that there exists a retraction of B onto its boundary, which we're in the process of proving cannot exist. Fictitious or not, we will need the result of this exercise in the next section!

4.3 Proof of the Brouwer Fixed-Point Theorem

We know now that the Brouwer Fixed-Point Theorem (henceforth "BFPT") is equivalent to the No-Retraction Theorem (henceforth "NRT") in the sense that each implies the other. In this section we'll show that the BFPT follows from a "C^1 version"
of NRT, which we'll then proceed to establish. Here is an outline of the argument.
First we'll show that:

$$C^1\text{-NRT} \implies C^1\text{-BFPT} \implies \text{BFPT} \qquad (*)$$

where, the prefix "C^1-" means that the result is being claimed only for maps whose
(real-valued) coordinate functions have continuous first order partial derivatives on
some open set that contains B. Then we'll get down to business and prove C^1-NRT.

(a) C^1-BFPT $\implies$ BFPT. The key is the following approximation theorem:

   Given $f: B \to \mathbb{R}^N$ continuous and $\varepsilon > 0$ there exists a C^1 map $g: B \to \mathbb{R}^N$
   with $|f(x) - g(x)| \leq \varepsilon$ for every $x \in B$.

See Appendix A.2 for a proof.[2] Now suppose $f: B \to B$ is a continuous map.
To show that f has a fixed point, let $\varepsilon > 0$ be given and choose—by the above-
mentioned approximation result—a C^1 map $f_\varepsilon: B \to \mathbb{R}^N$ with

$$|f_\varepsilon(x) - (1-\varepsilon)f(x)| \leq \varepsilon \qquad (x \in B). \qquad (4.5)$$

By the "reverse triangle inequality" we have $|f_\varepsilon(x)| - (1-\varepsilon)|f(x)| \leq \varepsilon$ for every
$x \in B$, i.e.,

$$|f_\varepsilon(x)| \leq \varepsilon + (1-\varepsilon)|f(x)| \leq \varepsilon + (1-\varepsilon) = 1.$$

Thus f_ε maps B into itself, so by our assumption that the C^1-BFPT holds, f_ε has a
fixed point $p_\varepsilon \in B$. By the (ordinary) triangle inequality, for every $x \in B$:

$$|f_\varepsilon(x) - f(x)| = |f_\varepsilon(x) - (1-\varepsilon)f(x) - \varepsilon f(x)|$$

$$\leq |f_\varepsilon(x) - (1-\varepsilon)f(x)| + \varepsilon|f(x)|$$

$$\leq \varepsilon + \varepsilon = 2\varepsilon$$

so in particular

$$|f(p_\varepsilon) - p_\varepsilon| = |f(p_\varepsilon) - f_\varepsilon(p_\varepsilon)| \leq 2\varepsilon \qquad (k \in \mathbb{N}), \qquad (4.6)$$

i.e., p_ε is a "2ε-approximate fixed point" for f. Since ε is an arbitrary positive
number, the Approximate-Fixed-Point Lemma (Lemma 2.2, page 24) guarantees
that f has a fixed point.   □

---

[2] More is true: the Stone–Weierstrass Theorem (see, e.g., [101, Theorem 7.6, page 159]) guarantees
that the coordinate functions of g can even be chosen to be polynomials (in n variables).

(b) C^1-*NRT* $\implies$ C^1-*BFPT*. Suppose C^1-BFPT fails, so there exists a C^1 map $f: B \to B$ with no fixed point. We'll show that in this case the retraction P given by Eqs. (4.1)–(4.4) on page 43 is C^1 on B. In the defining Eq. (4.2) for the unit vector u, the function $x - f(x)$ is C^1 and never zero, hence the denominator $|x - f(x)|$ is C^1 and (thanks to the compactness of B) bounded away from zero on B. Thus u is C^1 on B. The only issue left is the C^1 nature of the parameter $\lambda(x)$ on the right-hand side of Eq. (4.1), but this follows immediately from Eq. (4.4) and the fact that, on the right-hand side of that equation, the quantity under the radical sign is C^1 and—thanks to Exercise 4.2—strictly positive for each $x \in B$. □

(c) *Proof of C^1-NRT.* This is the heart of our proof of the BFPT. Suppose C^1-NRT is false, i.e., suppose there exists a C^1 retraction P taking B onto its boundary. We will show that this leads to a contradiction. The argument takes place in several steps.

STEP I: *A bridge from the identity map to P.*
    For $0 \le t \le 1$ define the map $P_t: B \to \mathbb{R}^N$ by

$$P_t(x) = (1-t)x + tP(x) \qquad (x \in B). \tag{4.7}$$

Directly from this definition it follows that:

(a) P_0 is the identity map on B, while $P_1 = P$.
(b)   Each P_t is a C^1 map that—since each of its values is a convex combination of two elements of B—takes B into itself.
(c)   Each map P_t fixes every point of ∂B.

For the next step let B° denote the interior of B, i.e., the open unit ball of $\mathbb{R}^N$.

STEP II: *There exists $t_0 \in (0,1]$ such that for all $t \in [0,t_0]$,*

(a) $\det P_t'(x) > 0$ *for all $x \in B$.*
(b) P_t *is a homeomorphism of B° onto itself.*

Here $P_t'(x)$ is the derivative of P_t evaluated at $x \in B^\circ$ (see Appendix A.1); we view $P_t'(x)$ as an $N \times N$ matrix whose entries are continuous, real-valued functions on some open set that contains B. We're claiming that for t sufficiently close to zero, P_t inherits the salient properties of the identity map P_0. Let's defer the proof of this statement until we've seen how it leads to the desired contradiction.

STEP III: *Deriving the contradiction.* Define $h: [0,1] \to \mathbb{R}$ by the multiple Riemann integral:

$$h(t) = \int_{B^\circ} \det P_t'(x)\,dx \qquad (0 \le t \le 1).$$

By STEP II and the Change-of-Variable Theorem (Theorem A.4):

$$h(t) = \int_{P_t(B^\circ)} dx = \text{Volume of } B^\circ \qquad (0 \le t \le t_0). \tag{4.8}$$

Now $\det P_t'$ is a polynomial in t with continuous real-valued coefficients, so by (4.8) $h(t)$ is a polynomial in t that, on the interval $[0, t_0]$, takes the constant value "volume of $B°$," and so has that constant value for *all* $t \in [0, 1]$. In particular, $h(1) > 0$. But we're assuming that $P_1 = P$ maps $B°$ into the unit sphere ∂B, a subset of $\mathbb{R}^N$ that has no interior, so by the Inverse-Function Theorem (Appendix A, Theorem A.3) its derivative matrix $P'(x)$ is singular for every $x \in B°$. Thus for $t = 1$ the integrand on the right-hand side of (4.8) is identically zero, i.e., $h(1) = 0$. Contradiction!

PROOF OF STEP II. This takes place in several stages, each of which expresses the fact that as we restrict t to increasingly smaller values, P_t inherits successively more properties of the identity map P_0.

STEP IIA: *For all t sufficiently small, P_t is a homeomorphism of B onto $P_t(B)$.*

Because P is a C^1 map on the compact set B, the Mean-Value Inequality (Appendix A, Theorem A.2, page 184) provides a positive constant L such for each pair x, y of points in B,

$$|P(x) - P(y)| \leq L|x - y|,$$

i.e., P satisfies a "Lipschitz condition" on B. Thus for $x, y \in B$ and $0 \leq t \leq 1$:

$$|P_t(x) - P_t(y)| = |(1-t)(x-y) + t[P(x) - P(y)]|$$
$$\geq (1-t)|x-y| - t|P(x) - P(y)|$$
$$\geq (1-t)|x-y| - tL|x-y|$$
$$= [1 - t(1+L)]|x-y|.$$

Conclusion: For $0 \leq t < 1/(1+L)$ the mapping P_t takes B one-to-one into itself, and $(P_t)^{-1}$ satisfies a Lipschitz condition, hence is continuous. In other words, for all sufficiently small t, the mapping P_t is a homeomorphism taking B onto some subset of B.

   Our goal now is to show that, at least for t sufficiently small, this subset is all of B. Since P_t is the identity map on ∂B, it will be enough to show that $P_t(B°) = B°$ for all sufficiently small t.

STEP IIB: *For all t sufficiently small, $P_t(B°)$ is an open subset of $B°$.*

From the definition (4.7) of P_t we see that for each $t \in [0, 1]$:

$$P_t'(x) = (1-t)I + tP'(x) \qquad (x \in B)$$

where I denotes the $N \times N$ identity matrix. Thus the "C^1-ness" of the retraction P translates into continuity for the map $(t, x) \rightarrow P_t'(x)$ as it takes the compact product space $[0, 1] \times B$ into the space of $N \times N$ real matrices endowed with the metric of $\mathbb{R}^{N^2}$. Since continuous functions on compact metric spaces are uniformly continuous, the function $(t, x) \rightarrow \det P_t'(x)$ is a uniformly continuous real-valued function on $[0, 1] \times B$. Since $P_0'(x)$ is the $N \times N$ identity matrix for each $x \in B$ this uniform continuity implies (exercise) that there exists $0 < t_0 < 1/C$ (C being the constant of Step IIa) such that $\det P_t'(x) \geq 1/2$ for each $(t,x) \in [0,t_0] \times B$. This justifies our application of the change-of-variable formula in STEP III, and shows that $P_t'(x)$ is invertible for all $t \in [0,t_0]$ and all $x \in B$. Thus if $0 \leq t \leq t_0$ the Inverse-Function Theorem (Appendix A, Theorem A.3) shows that P_t maps open sets to open sets; in particular $P_t(B^\circ)$ is open in B°.

STEP IIC: *For all t as promised by* STEP IIB, $P_t(B^\circ) = B^\circ$.

Fix such a t, so P_t is a homeomorphism of B^0 onto $P_t(B^0)$. Suppose $P_t(B^\circ) \neq B^\circ$. Then there is a point $y_0 \in B^\circ$ that belongs to the boundary of $P_t(B^\circ)$. One can therefore choose a sequence (y_k) of points in $P_t(B^\circ)$ with $y_k \to y_0$. Thus there exists a sequence (x_k) in B° with $P_t(x_k) = y_k$ for each index k. Thanks to the compactness of B we may assume, upon replacing (x_k) by an appropriate subsequence, that $\lim_k x_k = x_0 \in B$. Thus $y_0 = P_t(x_0)$ by the continuity of P_t. It follows that $x_0 \in B^\circ$; otherwise x_0 would belong to ∂B so, because P_t is the identity map on ∂B, the point $y_0 = P_t(x_0)$ would equal x_0, and so would lie on ∂B, contradicting our assumption that y_0 lies in B°.

This completes the proof of STEP II, and with it, the proof of the Brouwer Fixed-Point Theorem.   □

4.4 Retraction and Convexity

So far the work of this chapter has concentrated on the equivalence of the Brouwer Fixed-Point Theorem and the No-Retraction Theorem. Here is a different (and very useful) connection between fixed points and retractions.

Theorem 4.3. *Every retract of a space with the fixed-point property has the fixed-point property.*

Proof. Suppose S is a metric space with the fixed-point property, A is a subset of S, and $P\colon S \to A$ is a retraction of S onto A. Let $f\colon A \to A$ be a continuous map. We need to show that f has a fixed point. Since $g = f \circ P$ maps S into itself it has a fixed point. Since g maps S into A this fixed point, call it a, belongs to A. But the restriction of P to A is the identity map, so $a = g(a) = f(P(a)) = f(a)$.   □

Which spaces have the fixed-point property? Every one-point space has it (trivially), and for each positive integer N the closed unit ball of $\mathbb{R}^N$ has it (The Brouwer Fixed-Point Theorem). No circle has it (nontrivial rotations have no fixed point), hence no closed curve (homeomorphic image of a circle) has it.

The extension of Brouwer's theorem provided by Theorem 4.3 allows us to exhibit more examples. Here is one that is "one dimensional," but not homeomorphic to a closed interval (exercise).

Example 4.4. The letter "X" has the fixed-point property.

Proof. Here "the letter 'X'" is the union in $\mathbb{R}^2$ of those parts of the lines $y = x$ and $y = -x$ that lie in B, the closed unit ball of $\mathbb{R}^2$ (a.k.a "the closed unit disc"). Then $X \subset B$, so by Brouwer's theorem and Theorem 4.3 above we need only show that X is a retract of B. We'll accomplish this by modifying the "non-orthogonal" projection introduced above in Sect. 4.1. The set X divides B into four quadrants, each bisected by the coordinate half-axes. Project each point in B onto X by moving it parallel to the closest coordinate axis. Thus, each point of a coordinate axis goes to the origin, each point of X stays fixed, each point of the region above X goes straight down onto X, each point to the right of X goes horizontally onto X, etc. The result is a map P that takes B onto X, and whose restriction to X is the identity. I leave it to you to convince yourself that P is continuous.   □

Exercise 4.3. Which capital letters of the English alphabet have the fixed-point property?

So much for amusing examples. Here's the result we're really after.

Theorem 4.5 (The "Convex" Brouwer Fixed-Point Theorem). *Every compact convex subset of $\mathbb{R}^N$ has the fixed-point property.*

Proof. Let C be a compact convex subset of $\mathbb{R}^N$.

Claim. C is a retract of $\mathbb{R}^N$.

Even though $\mathbb{R}^N$ does not have the fixed-point property, this will prove our result. Indeed, since C is compact it is contained in a closed ball B (not necessarily the unit ball now) which, by Brouwer's theorem, has the fixed-point property. The Claim will give us a retraction P of $\mathbb{R}^N$ onto C, and the restriction of P to B will be a retraction of B onto C. The result will then follow from Theorem 4.3.

Proof of the Claim. The retraction we're about to produce—important in its own right—is the *Closest-Point Map.* Suppose $x \in \mathbb{R}^N$. Since C is compact there is at least one point $\kappa \in C$ with $|x - \kappa| = \inf\{|x - c| : c \in C\}$ (Proof: There is a sequence (c_j) of points in C for which $|x - c_j|$ converges to the infimum in question. By the compactness of C, this sequence has a convergent subsequence, whose limit κ is a point that achieves the infimum).

The convexity of C guarantees that κ is the *unique* closest point in C to x. To prove this, suppose $k \in C$ is another point "closest to x." For convenience let

$$d = \inf\{|x - c| : c \in C\} = |x - \kappa| = |x - k|$$

Let $v = x - \kappa$ and $w = x - k$. By the Parallelogram Law:

$$|v + w|^2 + |v - w|^2 = 2|v|^2 + 2|w|^2 = 4d^2.$$

On the other hand, the convexity of C guarantees that $(\kappa + k)/2 \in C$, hence

$$|(v + w)/2| = |(\kappa + k)/2 - x| \geq d.$$

The last two displays yield

$$4d^2 + |v - w|^2 \leq 4d^2$$

so $0 = |v - w| = |k - \kappa|$, hence $\kappa = k$, as desired.

Now that we know there's a unique closest point in C to x, let's give it a name: $P(x)$. Thus P maps $\mathbb{R}^N$ onto C, and fixes each point of C. To show that P retracts $\mathbb{R}^N$ onto C we need only verify its continuity. The result below shows that this follows from the "closest-point uniqueness" from which the mapping P owes its definition.   □

Proposition 4.6. *Suppose (X,d) is a metric space and A is a compact subset of X such that every $x \in X$ has a unique closest point $P(x)$ in A. Then P is a retraction of X onto A.*

Proof. Define the function "distance to A" by

$$d_A(x) = \inf\{a \in A : d(x,a)\} \qquad (x \in X).$$

Note first that $d_A : X \to [0,\infty)$ is continuous; in fact, it is "non-expansive" in the sense that

$$|d_A(x) - d_A(y)| \leq d(x,y) \qquad (x,y \in X). \tag{4.9}$$

In fact this is true for every $A \subset X$. To see why, fix x and y in X; suppose (without loss of generality) that $d_A(x) \geq d_A(y)$. Then $d_A(x) \leq d(x,a) \leq d(x,y) + d(y,a)$ for every $a \in A$, from which follows (thanks to the fact that a was an arbitrary element of A) that $d_A(x) \leq d(x,y) + d_A(y)$, which is another way of stating (4.9).

Now let's return to our compact subset A that *does* have the "unique closest point" property, and the map $P(x) = $ "closest point in A to x." We're trying to show that P is continuous, so fix $x_0 \in X$ and suppose (x_n) is a sequence in X that converges to x_0. Our goal is to show that $P(x_n) \to P(x_0)$. Since A is compact, the sequence $(P(x_n))$ of closest points has a subsequence convergent to a point—call it y_0—of A. To keep notation under control, let's replace (temporarily) the whole sequence by this subsequence, so that $P(x_n) \to y_0$. Then:

$$d_A(x_0) = \lim_n d_A(x_n) \qquad \text{(continuity of } d_A)$$

$$= \lim_n d(x_n, P(x_n)) \qquad \text{(definition of } P)$$

$$= d(x_0, y_0) \qquad \text{(definition of } y_0)$$

so y_0 is a closest point in A to x_0, hence by uniqueness, $y_0 = P(x_0)$. This argument actually proves that if x_0 is a point of X and (x_n) is a sequence that converges to x_0, then every subsequence of (x_n) has a further subsequence whose image under P converges to $P(x_0)$. Thus $P(x_n) \to P(x_0)$, as desired. This completes the proof of the Proposition, and with it the proof that the closest-point mapping of $\mathbb{R}^N$ onto the compact convex subset C is continuous, hence a retraction.   □

Notes

Proof of the Brouwer Fixed-Point Theorem. The argument given here is C.A. Rogers' modification [99] of an argument due to John Milnor [78]. In [67] Peter Lax proves an "oriented" version of the change-of-variable formula for multiple integrals, and uses this result to provide a more direct proof of Brouwer's Theorem. For a differential-forms interpretation of Lax's change-of-variable argument see [53], which also gives a valuable survey of papers that offer analytic proofs of Brouwer's Theorem.

More on proofs of the Brouwer Fixed-Point Theorem. We've seen two proofs of the Brouwer Fixed-Point Theorem: the one in Chap. 2 (for $N = 2$) based on the Sperner Lemma, and the one in this chapter. There are many others; see [112] for a nice survey. Brouwer's original proof [18], published in 1912, used methods of (what has since become known as) algebraic topology. Simultaneously, and for the rest of his life, Brouwer thought deeply about the foundations of mathematics—a pursuit that ultimately led him, 40 years later, to renounce this proof of his theorem [19].

The "Closest Point Property" of convex sets. With a little more care we can weaken the compactness hypothesis on the convex set C to just "closed-ness." The idea is that an application, similar to the one above, of the Parallelogram Law shows that the "minimizing sequence" (c_j) discussed above is actually a Cauchy sequence, and therefore converges, its limit being the unique closest point in C to x. In case C is a linear subspace of $\mathbb{R}^N$ this closest point turns out to be the orthogonal projection of x onto C. These arguments generalize, with no essential changes, to the setting of infinite dimensional Hilbert space (see [125, Sect. 3.2, p. 26 ff.], for example).

Non-expansiveness of the closest-point map. For a closed convex subset of $\mathbb{R}^N$ (or more generally of a Hilbert space) the "closest point map" P is more than just continuous: it is "non-expansive" in the sense that

$$|P(x) - P(y)| \leq |x - y|$$

for all $x, y \in \mathbb{R}^N$; see, for example, [50, Theorem 3.13, p. 118] for the details.

Chapter 5
Nash Equilibrium

FIXED POINTS IN GAME THEORY

Overview. In this chapter we'll study John F. Nash's fundamental notion of "equilibrium" in game theory. Following Nash, we'll use the Brouwer Fixed-Point Theorem to prove the result at the heart of his 1994 Nobel Prize in Economics: *For every finite non-cooperative game, the mixed-strategy extension has a Nash Equilibrium.* All will be explained; no prior experience with game theory will be assumed.

Prerequisites. Basic linear algebra and set theory.

5.1 Mathematicians Go to Dinner

Here's an example that illustrates much of what is to follow. Twenty mathematicians go out to dinner, agreeing to split the bill equally. Upon arrival they discover that the restaurant offers only two choices: the $10 dinner and the $20 one. Nobody wants to spend $20. Assuming that each person acts only on the basis of "rational self-interest," wouldn't it be best for everyone to order the $10 dinner? Guess again!

On the basis of pure rational self-interest, each player reasons: "Everyone else is going to choose the $10 dinner, so if I choose the $20 one, I get it for just $10.50." Unfortunately everyone else has the same thought, so it seems most likely that—even though nobody wants to pay $20—everyone will choose the $20 dinner.

This is an example of *Nash Equilibrium.* Later we'll formalize the idea but for now let's consider a few more examples, all of which involve—instead of 20 players—just two.

© Springer International Publishing Switzerland 2016
J.H. Shapiro, *A Fixed-Point Farrago*, Universitext,
DOI 10.1007/978-3-319-27978-7_5

5.2 Four Examples

Here are four simple two-person games, each player's goal being: "maximize my payoff." At each play of the game, each player knows all the strategies and all the payoffs, but *not* the strategy the other will use.

Matching Pennies. Each player puts down a penny. If both coins show the same face (heads–heads or tails–tails), Player I wins the other's coin. If the faces differ (heads–tails or tails–heads), then Player II wins Player I's coin. For this game, and each example to follow, it's convenient to represent the situation by a "payoff matrix" whose entries represent the payoffs to each competitor for each possible play of the game. For Matching Pennies the payoff matrix for Player I is

$$
\begin{array}{cc}
\text{I}\backslash\text{II} & \begin{array}{cc} H & T \end{array} \\
\begin{array}{c} H \\ T \end{array} & \left[\begin{array}{cc} 1 & -1 \\ -1 & 1 \end{array} \right]
\end{array}
$$

Here the "(i,j)-entry" of the matrix represents Player I's payoff upon playing face i (H or T) to Player II's face j. For this game the payoff matrix for Player II is the negative of the one for Player I, i.e., Matching Pennies is a "zero-sum" game: each player's gain is the other's loss. In a two-person game it's often convenient to display the payoffs for both players in a "bi-matrix." For Matching Pennies this is

$$
\begin{array}{cc}
\text{I}\backslash\text{II} & \begin{array}{cc} H & \qquad T \end{array} \\
\begin{array}{c} H \\ T \end{array} & \left[\begin{array}{cc} (1,-1) & (-1,1) \\ (-1,1) & (1,-1) \end{array} \right]
\end{array}
$$

where now the entry at row i and column j ("I plays strategy i and II plays j") displays Player I's payoff in the first coordinate and Player II's in the second.

Rock–Paper–Scissors. Here each competitor has strategies: "rock," "paper," or "scissors." At each play of the game, the players use the familiar hand signals to simultaneously indicate their chosen strategies with the winner is determined by the rules: "paper covers rock," "scissors cuts paper," and "rock breaks scissors." Suppose that at each play of the game the winner receives one penny from the loser, with neither winning anything if both opt for the same strategy. The situation is captured by the 3×3 bi-matrix shown below, its "(i,j)-entry" representing both players' payoffs when Player I plays strategy $i \in \{r,p,s\}$ to Player II's strategy j.

$$
\begin{array}{cc}
\text{I}\backslash\text{II} & \begin{array}{ccc} r & \quad p & \quad s \end{array} \\
\begin{array}{c} r \\ p \\ s \end{array} & \left[\begin{array}{ccc} (0,0) & (-1,1) & (1,-1) \\ (1,-1) & (0,0) & (-1,1) \\ (-1,1) & (1,-1) & (0,0) \end{array} \right]
\end{array}
$$

Prisoners' Dilemma. This game imagines both players to be prisoners held in separate interrogation cells. The police are sure the suspects have committed a serious crime which, if this can be proven, will land each in prison for 5 years. However there's only enough evidence to convict the pair of a less serious infraction that carries a prison term of just 1 year. The prisoners are offered this deal: If one of them defects by implicating the other in the more serious crime while the other prisoner does *not* defect, then the defector will go free while the loyal one will get the maximum sentence. If each prisoner defects, then for cooperating each will receive a somewhat reduced sentence of 3 years. The bi-matrix below summarizes the payoffs, where "L" = "stay loyal" and "D" = "defect."

$$
\begin{array}{c}
\text{I}\backslash\text{II} \quad\quad L \quad\quad\quad\quad D \\
\begin{array}{c} L \\ D \end{array}
\left[
\begin{array}{cc}
(-1,-1) & (-5,0) \\
(0,-5) & (-3,-3)
\end{array}
\right]
\end{array}
$$

Battle of the Sexes. In this game Players I and II are a couple who wish to spend an evening out. Player I wants to go to the ball game (*B*) while Player II prefers the symphony (*S*). They work on opposite sides of town and plan to decide that afternoon, via some form of electronic communication, which event to attend. But a massive solar flare renders communication impossible, so each must guess the other's intention. The payoff matrix below indicates their preferences: the diagonal terms indicate that for each player the favored event is twice as desirable as the alternative while the cross-diagonals show that neither choice is so desirable as to warrant going alone.

$$
\begin{array}{c}
\text{I}\backslash\text{II} \quad\quad B \quad\quad\quad S \\
\begin{array}{c} B \\ S \end{array}
\left[
\begin{array}{cc}
(2,1) & (0,0) \\
(0,0) & (1,2)
\end{array}
\right]
\end{array}
$$

5.3 Nash Equilibrium

In Prisoners' Dilemma it appears at first glance that both players should remain loyal, thus insuring that—although nobody goes free—each gets a relatively light sentence. However if one player believes in the other's unswerving loyalty, then that player will do better by defecting. A quick look at the rows of the payoff matrix shows that Player I's strategy D *dominates* L in the sense that no matter which strategy Player II chooses, Player I's payoff for D is better than for L. Similarly, looking at the columns of the matrix, we see that Player II's strategy D also dominates L.

Thus each player has D as a dominant strategy, so—even though the best overall result would be for both to stay loyal—the safest is for each to defect.

Such "dominance" is not always present; neither of the other three games described above exhibits it. However in Battle of the Sexes the strategy pairs (B,B) and (S,S) share a salient feature of dominant pairs: If a player deviates unilaterally from either pair then that player's payoff decreases. The same is true of in the "Mathematicians go to Dinner" game for the strategy 20-tuple: "Everyone chooses the $20 dinner."

In the middle of the last century John F. Nash made a profound study of this "weaker-than-dominance" notion. Now called "Nash Equilibrium," it has become a cornerstone of modern economic theory. To describe it properly we need to formalize our notion of "non-cooperative game."

Definition 5.1. A *non-cooperative game* (henceforth, just a "game") consists of a set $\mathscr{P}$ of *players*, where each player P has

(a)   A *strategy set* Σ_P, and
(b)   A real-valued function u_P defined on the cartesian product $\Sigma = \prod_{P \in \mathscr{P}} \Sigma_P$ of strategy sets; this is player P's *payoff function*.

Each element $\sigma \in \Sigma$ represents a particular play of the game; think of it as a "vector" with coordinates indexed by the players, $\sigma(P)$ being the strategy chosen by Player P. Then $u_P(\sigma)$ is Player P's payoff for that particular play of the game. Upon defining $\mathscr{U} = \{u_P : P \in \mathscr{P}\}$, we can refer to the game described above as simply the triple $(\mathscr{P}, \Sigma, \mathscr{U})$.

In the examples of Sect. 5.2 the set $\mathscr{P}$ of players has just has two elements, while the strategy sets are finite and the same for each player. In Rock–Paper–Scissors, for example, $\mathscr{P} = \{\text{Player I}, \text{Player II}\}$, the strategy sets are $\Sigma_I = \Sigma_{II} = \{r, p, s\}$, the collection Σ of strategy "vectors" is the nine-element cartesian product

$$\Sigma_I \times \Sigma_{II} = \{(x,y) : x, y \in \{r, p, s\}\},$$

and the payoff function for Player I is summarized by the matrix below; $u_I(x,y)$ being the number located at the intersection of the x-row and the y-column

$$
\begin{array}{cc}
\text{I\textbackslash II} & \begin{array}{ccc} r & p & s \end{array} \\
\begin{array}{c} r \\ p \\ s \end{array} &
\left[\begin{array}{ccc}
0 & -1 & 1 \\
1 & 0 & -1 \\
-1 & 1 & 0
\end{array} \right]
\end{array}
$$

(that of Player II is represented by the negative of this matrix; like Matching Pennies, this is a "zero-sum" game). Thus, for example, $u_I(p,s) = -1 = -u_{II}(p,s)$.

For "Mathematicians go to dinner" our player set is $\mathscr{P} = \{P_1, P_2, \ldots, P_{20}\}$, each player P_j having the same strategy set $\Sigma_j = \{10, 20\}$ $(1 \leq j \leq 20)$. Thus the set of strategy vectors $\Sigma = \prod_{j=1}^{20} \Sigma_j$ is the set of 20-tuples, each coordinate of which is either 10 or 20. The description of the game suggests that Player P_j's payoff for strategy vector $x = (x_1, x_2, \ldots, x_{20})$ should be

$$u_j(x) = x_j - \frac{1}{20} \sum_{k=1}^{20} x_k \qquad (j = 1, 2, \ldots, 20),$$

where on the right-hand side the first term is the value of P_j's dinner, and the second is the amount P_j pays for it: the total cost of all 20 dinners, averaged over the participants.

Definition 5.2 (Unilateral Change of Strategy). In a non-cooperative game $(\mathscr{P}, \Sigma, \mathscr{U})$, suppose $P \in \mathscr{P}$ and σ and σ' are strategy vectors that differ only in the P-coordinate (i.e., $\sigma(Q) = \sigma'(Q)$ for all $Q \in \mathscr{P} \setminus \{P\}$, and $\sigma'(P) \neq \sigma(P)$). In this case we'll say that for Player P: σ' is a *unilateral change of strategy from* σ.

Definition 5.3 (Nash Equilibrium). For a non-cooperative game $(\mathscr{P}, \Sigma, \mathscr{U})$: a *Nash equilibrium* is a strategy vector σ^* with the property that: For each player, no unilateral change of strategy from σ^* results in a strictly better payoff. More precisely: "$\sigma^* \in \Sigma$ is a Nash equilibrium" means

$$P \in \mathscr{P}, \ \sigma \in \Sigma, \text{ and } \sigma(Q) = \sigma^*(Q) \ \forall Q \in \mathscr{P} \setminus \{P\} \implies u_P(\sigma) \leq u_P(\sigma^*).$$

Nash equilibrium is perhaps best interpreted in terms of "best response." To avoid notational complications let's consider this notion just for two-person games. In this setting Player I and Player II have strategy sets X and Y, respectively, and $\Sigma = X \times Y$: the set of ordered pairs (x, y), where x ranges through X and y through Y.

Definition 5.4 (Best Response). To say that Player I's strategy $x^* \in X$ is a *best response* to Player II's strategy $y \in Y$ means that $u_I(x^*, y) = \max_{x \in X} u_I(x, y)$.

We'll always assume the maximum in this definition exists, either because the strategy set available to each player is finite, or as in the next section, because it is a compact subset of $\mathbb{R}^N$ with the payoff functions continuous in each variable.

Similar language defines Player II's best response to a given strategy of Player I. With this terminology:

> For a two-person game: a Nash Equilibrium is a pair of strategies, each of which is a best response to the other.

In terms of bi-matrices it's easy to identify such mutual best-response strategy pairs, should they exist. For each column of the matrix, underline the entry or entries for which Player I's payoff is largest (the "best response(s)" of Player I to Player II's strategy for that column). Similarly, for each row, place a line *over* each entry for which Player II's payoff is largest. If an entry is marked twice, it is a Nash equilibrium. It's easy to check this way that (D, D) is a Nash equilibrium for Prisoners' Dilemma, that Rock–Paper–Scissors and Matching Pennies have no Nash equilibrium, and that Battle of the Sexes has *two* Nash equilibria: (B, B) and (S, S). Thus:

(a) Not every game has a Nash equilibrium (Matching Pennies, Rock–Paper–Scissors).
(b) If it exists, Nash equilibrium need not be unique (Battle of the Sexes).
(c) If it exists, Nash equilibrium need not provide the best possible outcome for any player (Mathematicians go to Dinner, Prisoners' Dilemma).

5.4 Mixed Strategies

Let's return to Matching Pennies. In a sequence of consecutive plays, how can each competitor guard against falling into a behavior pattern discoverable by the other? Each could choose heads or tails *randomly*, for example, tossing the coin rather than just putting it down. Their coins might be biased; say Player I chooses a coin that has probability $p \in [0,1]$ of coming up Heads (and $1-p$ of Tails), while Player II opts for one that has probability $q \in [0,1]$ of Heads. So now we have the makings of a new game, with the same players, but with new ("mixed") strategies: *probability distributions* over the original ("pure") strategies. Let's represent the mixed strategy "Heads with probability p and Tails with probability $1-p$" by the vector

$$x_p = (p, 1-p) = pH + (1-p)T$$

where now $H = (1,0)$ denotes the pure strategy of choosing Heads with probability 1, and $T = (0,1)$ the pure strategy of choosing Tails with probability 1. The strategy set for each player can now be visualized as the line segment in $\mathbb{R}^2$ joining $H = (1,0)$ to $T = (0,1)$.

We need payoff functions for this new game, and for these we choose *expected payoffs*, calculated in the obvious way: Player I's expected payoff for playing the pure-strategy Heads to Player II's mixed strategy $y_q = qH + (1-q)T$ is[1]

$$u_I(H, y_q) = 1 \cdot q + (-1) \cdot (1-q) = 2q - 1,$$

while for playing Tails to Player II's y_q it is

$$u_I(T, y_q) = -1 \cdot q + 1 \cdot (1-q) = 1 - 2q.$$

Thus Player I's expected payoff for playing strategy x_p to Player II's y_q is

$$u_I(x_p, y_q) = p \cdot u_I(H, y_q) + (1-p) \cdot u_I(T, y_q) = (2p-1)(2q-1). \qquad (5.1)$$

Similarly, Player II's expected payoff for playing strategy y_q to Player I's x_p is

$$u_{II}(x_p, y_q) = (1-2q)(2p-1) = -u_I(x_p, y_q). \qquad (5.2)$$

In this way we arrive at a new game, having the same players but a larger strategy set to which the original payoff functions have been extended. We'll call this new game the *mixed-strategy extension* of the original one; note that by Eq. (5.2) the new game preserves the zero-sum nature of the original.

Recall that the original Matching-Pennies game had no Nash equilibrium. However for the mixed-strategy extension *there is one*, namely the strategy pair $(x_{1/2}, y_{1/2})$ wherein each player chooses to flip a fair coin. The (expected) payoffs for this strategy pair are not spectacular—zero for each—but neither player's payoff

---

[1] Even though y_q and x_q are the same vectors in $\mathbb{R}^2$, it's useful to pretend that the strategy sets of the two players are different, and denote Player I's mixed strategies by x's and Player II's by y's.

can be unilaterally improved. For example, if Player II chooses to play $y_{1/2}$ then, according to (5.1) above, Player I's expected payoff is $u_I(x_p, y_{1/2}) = 0$ *for every* $p \in [0,1]$, and similarly for Player II when Player I chooses to play $x_{1/2}$. In other words, each strategy x_p with $(0 \leq p \leq 1)$ is a *best response* for Player I to Player II's fair-coin strategy $y_{1/2}$, and vice versa. Thus each strategy in the pair $(x_{1/2}, y_{1/2})$ is a best response to the other, i.e., that strategy pair is a Nash equilibrium.

The "indifference of best response" that surfaced in the last paragraph is not an accident. It arises from the linearity of mixed-strategy payoffs in each variable, and will be important in Sect. 5.7 where we prove Nash's general theorem on the existence of equilibria. Right now, however, let's show that the Nash equilibrium we just found is the only one possible for the mixed-strategy extension of Matching Pennies.

Proposition 5.5. *For the mixed-strategy extension of Matching Pennies, the "fair-coin" strategy pair $(x_{1/2}, y_{1/2})$ is the only Nash equilibrium.*

Proof. Suppose (x_{p^*}, y_{q^*}) were a Nash equilibrium with $q^* \neq 1/2$. Then by definition we'd have $u_I(x_{p^*}, y_{q^*}) \geq u_I(x_p, y_{q^*})$ for each $p \in [0,1]$. Now substitute into the right-hand side of this inequality: $p = 1$ if $q^* > 1/2$ (i.e., Player I should play Heads if Player II is more likely to play Heads), and $p = 0$ if $q^* < 1/2$. By Eq. (5.1) Player I's payoff in either case would be $|2q - 1| > 0$. Thus $u_I(x_{p^*}, y_{q^*}) > 0$. This, along with (5.1), would imply that $p^* \neq 1/2$, in which case the same argument would show that $u_{II}(x_{p^*}, y_{q^*}) > 0$, contradicting the fact that the mixed-strategy extension of Matching Pennies is a zero-sum game. □

> *Exercise* 5.1. Compute the payoff functions for the mixed-strategy extension of Prisoners' Dilemma, and show that "D" is still each player's best response to every mixed strategy of the other. Thus (D,D) is still the only Nash equilibrium.

5.5 The Mixed-Strategy Extension of Rock–Paper–Scissors

Recall that for the Rock–Paper–Scissors game the payoff matrix for Player I is (omitting the labels)

$$A = \begin{bmatrix} 0 & -1 & 1 \\ 1 & 0 & -1 \\ -1 & 1 & 0 \end{bmatrix},$$

and the corresponding matrix for Player II is just $-A$; it's a zero-sum game.

Suppose that, in the mixed-strategy extension of this game, Player II uses the strategy $y = y_1 r + y_2 p + y_3 s$ ("rock" with probability y_1, ...), which we interpret as a probability vector (y_1, y_2, y_2) (non-negative components that sum to 1, i.e., a vector in the standard simplex Π_3). Then the pure-strategy payoffs to Player I can be expressed by matrix multiplication:

$$\begin{bmatrix} u_I(r,y) \\ u_I(p,y) \\ u_I(s,y) \end{bmatrix} = Ay^t = A \begin{bmatrix} y_1 \\ y_2 \\ y_3 \end{bmatrix} = \begin{bmatrix} -y_2 + y_3 \\ y_1 - y_3 \\ -y_1 + y_2 \end{bmatrix}. \tag{5.3}$$

Thus the mixed-strategy payoff to Player I upon playing strategy $x = x_1 r + x_2 p + x_3 s$ (where $x = (x_1, x_2, x_3)$ is a probability vector) is

$$u_I(x,y) = xAy^t = [x_1, x_2, x_3] \begin{bmatrix} -y_2 + y_3 \\ y_1 - y_3 \\ -y_1 + y_2 \end{bmatrix}$$

$$= (-y_2 + y_3)x_1 + (y_1 - y_3)x_2 + (-y_1 + y_2)x_3) \tag{5.4}$$

$$= (x_2 - x_3)y_1 + (-x_1 + x_3)y_2 + (x_1 - x_2)y_3.$$

We see from the second line of (5.4) that if Player II employs the "uniform" strategy $y^* = \frac{1}{3}r + \frac{1}{3}p + \frac{1}{3}s$, then *every* mixed strategy $x = x_1 r + x_2 p + x_3 s$ for Player I is a best response in that $u_I(x, y^*) = 0$ (another instance of the soon-to-be proved "Principle of Indifference"). In other words, a unilateral change of strategy cannot improve Player I's payoff. Since $u_{II} = -u_I$, the same is true for Player II if Player I uses the uniform strategy. Thus the strategy pair ("uniform","uniform") is a Nash equilibrium for the mixed-strategy extension of Rock–Paper–Scissors.

Proposition 5.6. *The mixed-strategy extension of Rock–Paper–Scissors has no other Nash equilibrium.*

Proof. The proof is similar to the uniqueness argument for Matching Pennies. Since Rock–Paper–Scissors is a zero-sum game, so is its mixed-strategy extension. Let (x^*, y^*) be a Nash equilibrium pair for this extended game. Suppose for the sake of contradiction that y^* is *not* the uniform strategy, i.e., some components are not $1/3$.

The matrix equation (5.3) tells us that for each strategy y chosen by Player II, the components of Player I's payoff vector (the left-hand side of (5.3)) sum to zero, and this vector is identically zero if and only if all the components of y are the same (i.e., $= 1/3$). Thus some component of the left-hand side of (5.3) must be positive, i.e., Player I has a pure strategy x for which $u_I(x, y^*) > 0$, hence $u_I(x^*, y^*) > 0$ (since x^* is a *best* response for Player I to Player II's strategy y^*). It follows from the second line of (5.4) that x^* is not the uniform strategy, so one of *its* components must be positive, hence by the same reasoning $u_{II}(x^*, y^*) > 0$. Thus both $u_I(x^*, y^*)$ and $u_{II}(x^*, y^*)$ are > 0, contradicting the fact that $u_I = -u_{II}$. □

5.6 The Principle of Indifference

The phenomenon of "indifference" noted in our analyses of Matching Pennies and Rock–Paper–Scissors will be crucial to our proof of Nash's existence theorem. Recall from Definition 1.7 (Sect. 1.7, page 11) the *standard simplex* Π_N in $\mathbb{R}^N$. This is the set of vectors in $\mathbb{R}^N$ with non-negative coordinates that sum to 1, i.e., the convex hull[2] of the standard unit vectors $\{e_1, e_2, \dots e_N\}$ for $\mathbb{R}^N$. Here we'll think of the elements of Π_N as *probability vectors*. The idea is that each standard unit vector represents a *pure strategy* in an N-person game, while the vector $(\xi_1, \xi_2, \dots, \xi_N) = \sum_j \xi_j e_j \in \Pi_N$ represents the *mixed strategy* of playing the strategy e_j with probability ξ_j.

Definition 5.7. The *support* of a vector $x \in \Pi_N$ is the set of vectors e_j in the representation $x = \sum_j \xi_j e_j$ for which $\xi_j \neq 0$. Notation: $\mathrm{spt}(x)$.

We begin with a simple maximum principle for restrictions to Π_N of linear functions $\mathbb{R}^N \to \mathbb{R}$. Each such function f (henceforth simply called a "linear function on Π_N") is continuous (Exercise 1.7, page 12), which—along with the compactness of Π_N (established in Exercise 1.4, page 12)—guarantees that f attains its maximum at some point of Π_N. This point has special properties.

Lemma 5.8. *Suppose $f \colon \Pi_N \to \mathbb{R}$ is linear and attains its maximum at $x^* \in \Pi_N$. Then $f \equiv f(x^*)$ on $\mathrm{spt}(x^*)$, and hence on the convex hull of that support.*

Proof. Let $m = f(x^*) = \max\{f(x) : x \in \Pi_N\}$ and let J denote the set of indices j for which $e_j \in \mathrm{spt}(x^*)$. Then

(a) $x^* = \sum_{j \in J} \xi_j e_j$, where $\xi_j > 0$ for each $j \in J$ and $\sum_{j \in J} \xi_j = 1$.
(b) $f(e_j) \leq m$ for each j.

Suppose for the sake of contradiction that $f(e_k) < m$ for some $e_k \in \mathrm{spt}(x^*)$. Then by the linearity of f:

$$m = f(x^*) = \sum_{j \in J} \xi_j f(e_j) = \sum_{j \in J \setminus \{k\}} \xi_j \underbrace{f(e_j)}_{\leq m} + \xi_k \underbrace{f(e_k)}_{< m} < m \sum_{j \in J} \xi_j = m,$$

hence $m < m$: contradiction. Thus $f(e_j) = m$ for every e_j in $\mathrm{spt}(x^*)$. Again by linearity, $f \equiv m$ on the convex hull of $\mathrm{spt}(x^*)$.   □

In addition to the continuity of f this proof used only the fact that f "respects convex combinations" in the sense that if $x_1, x_2, \dots, x_n \in \Pi_N$ and $(\lambda_j)_{j=1}^N \in \Pi_N$, then $f(\sum_j \lambda_j x_j) = \sum_j \lambda_j f(x_j)$. Functions on convex sets with this property are called *affine* (example: every function of the form "linear + constant").

Exercise 5.2. Suppose C is a convex subset of $\mathbb{R}^N$ (or more generally, of a real vector space), and f is an affine map taking C into $\mathbb{R}^M$ (or more generally, into a possibly different real vector space). Show that $f(C)$ is convex, and $f^{-1}(E)$ is convex for any convex subset E of $f(C)$.

---

[2] See Sect. C.1, Appendix C for the basic concepts involving convex sets.

Corollary 5.9 (The Principle of Indifference). *In the mixed-strategy extension of a two-person game: suppose x^* is a best response of Player I to Player II's strategy y. Then $u_I(\cdot, y)$ is constant on the convex hull of the support of x^*. If y^* is a best response of Player II to Player I's strategy x, then $u_{II}(x, \cdot)$ is constant on* conv spt (y^*).

Proof. We're given that x^* is a best response of Player I to Player II's strategy y. Then the function $u_I(\cdot, y)$, being linear in the first variable, attains its maximum at x^*. By Lemma 5.8 it's therefore constant on the convex hull of the support of x^*. The same argument works upon reversing the players' roles.   □

In our work on the mixed-strategy extensions of both Matching Pennies and Rock–Paper–Scissors we found that for each player: *no unilateral deviation from the equilibrium (x^*, y^*) could change that player's payoff.* Corollary 5.9 shows this to be a consequence of that fact that, in each example, spt $(x^*) = $ spt $(y^*) = $ all the pure strategies.

Example 5.10. Consider the mixed-strategy extension of "Battle of the Sexes." Let x_p denote Player I's mixed strategy $pB + (1-p)S$ and let y_q denote Player II's strategy $qB + (1-q)S$, where $0 \le p, q \le 1$. If Player II plays strategy y_q then Player I's pure-strategy responses result in these payoffs:

$$u_I(B, y_q) = qu_I(B, B) + (1-q)u_I(B, S) = 2q$$

and

$$u_I(S, y_q) = qu_I(S, B) + (1-q)u_I(S, S) = 1 - q.$$

These payoffs are equal precisely when $q = 1/3$, hence for each $0 \le p \le 1$:

$$u_I(B, y_{1/3}) = u_I(S, y_{1/3}) = u_I(x_p, y_{1/3}) = 2/3$$

Thus by linearity:

> Each of Player I's mixed strategies is a best response to Player II's strategy $y_{1/3} = \frac{1}{3}B + \frac{2}{3}S$. Furthermore, the strategy $y_{1/3}$ is the only strategy of Player II with this property.

Similar computations yield the fact that each of Player II's strategies is a best response to Player I's strategy $x_{2/3} = \frac{2}{3}B + \frac{1}{3}S$, and that $x_{2/3}$ is unique in this regard. Thus each coordinate of the strategy pair $(x_{2/3}, y_{1/3})$ is a best response to the other, and this is the only mixed-strategy pair for which this happens.

Conclusion: In addition to the two pure-strategy Nash equilibria (B, B) and (S, S) for the "Battle of the Sexes" game, there is a mixed-strategy equilibrium $(x_{2/3}, y_{1/3})$, and these three equilibria are the only ones possible.

5.7 Nash's Theorem

We're finally ready to state and prove Nash's famous theorem establishing the existence of Nash equilibria for mixed-strategy extensions of finite games. For simplicity we'll state and prove the theorem for two-player games, after which we'll point out how the argument applies as well to the general situation.

Theorem 5.11 (Nash 1950). *For every non-cooperative two-player game, the mixed-strategy extension has a Nash equilibrium.*

Proof. Suppose Player I has M pure strategies and Player II has N of them. Then, as we observed in Sect. 5.5 for the special case of "Rock–Paper–Scissors," the payoff functions for each player can be represented in terms of $M \times N$ payoff matrices A (for Player I) and B (for Player II):

$$u_I(x,y) = xAy^t \quad \text{and} \quad u_{II}(x,y) = xBy^t \tag{5.5}$$

where the probability vectors $x \in \Pi_M$ and $y \in \Pi_N$ are regarded as row matrices, and the superscript "t" means "transpose." Both payoff functions are, in each variable separately, linear and continuous.

Best Response. Since Π_N is a compact subset of $\mathbb{R}^N$, for each fixed $y \in \Pi_N$ the payoff function $u_I(\cdot, y)$ attains its maximum thereon, say at $x^* \in \Pi_M$. In the terminology of Definition 5.4, the strategy x^* is a *best response* for Player I to Player II's strategy y. By Lemma 5.8, every vector in the support of x^*, and indeed in the convex hull of this support, is also a best response to y. In particular, for each strategy $y \in \Pi_N$ of Player II, Player I has a *pure strategy* best response. Similarly, Player II has a pure-strategy best response to each strategy of Player I.

Measuring Improvement. Since Player I's pure strategies contain a best response to Player II's strategy $y \in \Pi_N$, we can measure how much a strategy $x \in \Pi_M$ differs from that best response by considering the non-negative functions δ_i defined by

$$\delta_i(x,y) = \max\{u_I(e_i, y) - u_I(x,y), 0\} \quad (x \in \Pi_M, y \in \Pi_N)$$

for $i = 1, 2, \ldots, M$. Let $\delta(x,y)$ be the vector in $\mathbb{R}^m$ whose i-th component is $\delta_i(x,y)$. A little thought will convince you that for Player I:

x is a best response to y if and only if $\delta(x,y) = 0$.

New strategies from old. Consider the map $T_I: \Pi_M \times \Pi_N \to \mathbb{R}^M$ defined by

$$T_I(x,y) = \frac{x + \delta(x,y)}{1 + \sum_k \delta_k(x,y)} \quad (x \in \Pi_M, y \in \Pi_N). \tag{5.6}$$

The coordinates of $T_I(x,y)$ are all non-negative and they sum to one, so T_I maps $\Pi_M \times \Pi_N$ (continuously) into Π_M.

Claim: $x \in \Pi_M$ is a best response to $y \in \Pi_N$ if and only if $T_I(x,y) = x$.

Proof of Claim. We've observed that x is a best response to y if and only if $\delta(x,y) = 0$, and from the definition of T_I we see that if $\delta(x,y) = 0$ then $T_I(x,y) = x$.

It remains to prove the converse, which we'll do in the contrapositive direction. To this end, suppose $x \in \Pi_M$ is *not* a best response to $y \in \Pi_N$. We wish to prove that $T_I(x,y) \neq x$. By Lemma 5.8, Player I has a pure strategy e_i (the i-th standard unit vector for $\mathbb{R}^M$) that *is* a best response to y. Because x *is not* such a best response, $\delta_i(x,y) > 0$ and $x \neq e_i$, hence the i-th coordinate of x is < 1. Thus $\sum_k \delta_k(x,y) > 0$ and $x \cdot \delta(x,y) < \sum_k \delta_k(x,y)$ (where "$\cdot$" denote the dot product of vectors in $\mathbb{R}^M$). It follows that

$$x \cdot T_1(x,y) = \frac{x \cdot x + x \cdot \delta(x,y)}{1 + \sum_k \delta_k(x,y)} = \frac{1 + x \cdot \delta(x,y)}{1 + \sum_k \delta_k(x,y)} < 1,$$

hence $T_1(x,y) \neq x$, and the Claim is proved.

Enter the Brouwer Fixed-Point Theorem. So far we've created a continuous map $T_I : \Pi_M \times \Pi_N \to \Pi_M$ with the property that $x \in \Pi_M$ is a best response for Player I to Player II's strategy $y \in \Pi_N$ if and only if $T_I(x,y) = x$. Similarly there's a continuous map $T_{II} : \Pi_M \times \Pi_N \to \Pi_N$ such that $y \in \Pi_N$ is Player II's best response to Player I's strategy $x \in \Pi_M$ if and only if $T_{II}(x,y) = y$. Finally, define $T : \Pi_M \times \Pi_N \to \Pi_M \times \Pi_N$ by

$$T(x,y) = (T_I(x,y), T_{II}(x,y)) \qquad (x,y) \in \Pi_M \times \Pi_N.$$

Then T is a continuous self-map of the compact, convex subset $\Pi_M \times \Pi_N$ of $\mathbb{R}^{MN}$, so by the *Brouwer Fixed-Point Theorem* it has a fixed point (x^*, y^*). Thus x^* and y^* are best responses to each other, i.e., the pair (x^*, y^*) is a Nash equilibrium. □

To apply this argument to N-player games, note that throughout, Player II's strategy y is "inert" in that it could as well have been the strategy vector for the other $N-1$ players in an N-player game. That observation having been made, it's a routine matter to extend the proof given above to the more general situation.

5.8 The Minimax Theorem

Let's begin with a simple observation about real-valued functions of two variables. Suppose X and Y are sets and $u : X \times Y \to \mathbb{R}$ is a real-valued function. Then (assuming for simplicity that all maxima and minima mentioned below exist) for each $x_0 \in X$ and $y_0 \in Y$ we have

$$\min_{y \in Y} u(x_0, y) \leq u(x_0, y_0) \leq \max_{x \in X} u(x, y_0)$$

whereupon

$$\max_{x \in X} \min_{y \in Y} u(x, y) \leq \min_{y \in Y} \max_{x \in X} u(x, y). \tag{5.7}$$

If we interpret X and Y as strategies for a two-player game, and u as the payoff function for the X-strategy player, then $\max_{x \in X} u(x, y_0)$ is that player's best payoff when the other's strategy is y_0, so the right-hand side of (5.7) is the X-player's "worst-best" payoff. Similarly, $\min_{y \in Y} u(x, y)$ is the X-player's worst payoff for playing strategy x, hence the left-hand side of (5.7) is that player's "best-worst" payoff. Thus inequality (5.7) says for the X-player that: "worst-best is always at least as good as best-worst."

Exercise 5.3. Show (e.g., by considering the example $u(x, y) = |x - y|$ for $0 \le x, y \le 1$) that "worst-best" can be strictly better than "best-worst."

For mixed-strategy extensions of finite, two-person, non-cooperative, zero-sum games: Nash's theorem shows that there is actually *equality* in (5.7). This is von Neumann's famous *Minimax Theorem*, which was considered, pre-Nash, to be the fundamental theorem of game theory.

Theorem 5.12 (The Minimax Theorem). *For every $M \times N$ real matrix A there exist vectors $x^* \in \Pi_M$ and $y^* \in \Pi_N$ such that:*

$$\max_{x \in \Pi_M} \min_{y \in \Pi_N} xAy^t = x^*Ay^{*t} = \min_{y \in \Pi_N} \max_{x \in \Pi_M} xAy^t \tag{5.8}$$

Proof. Consider the mixed-strategy extension of the two-person game where Player I's payoff matrix is A and Player II's is $-A$. Nash's theorem asserts that there exists a mixed-strategy pair $x^* \in \Pi_M$, $y^* \in \Pi_N$ such that

$$xAy^{*t} \le x^*Ay^{*t} \quad \forall x \in \Pi_M \quad \text{and} \quad x^*(-A)y^t \le x^*(-A)y^{*t} \quad \forall y \in \Pi_N,$$

i.e.,

$$xAy^{*t} \le x^*Ay^{*t} \le x^*Ay^t \qquad \forall (x, y) \in \Pi_M \times \Pi_N.$$

Thus for each $(x, y) \in \Pi_M \times \Pi_N$:

$$\max_{x \in \Pi_M} xAy^{*t} \le x^*Ay^{*t} \le \min_{y \in \Pi_N} x^*Ay^t,$$

from which it follows that

$$\min_{y \in \Pi_N} \max_{x \in \Pi_M} xAy^t \le x^*Ay^{*t} \le \max_{x \in \Pi_M} \min_{y \in \Pi_N} xAy^t.$$

This, along with (5.7) establishes (5.8).   □

The Minimax Theorem asserts that for every finite, two-person, zero-sum game there is a number V, and mixed strategies x^* for Player I and y^* for Player II, such that Player I's payoff for strategy pair (x^*, y^*) is V, Player II's payoff is $-V$, and neither player's payoff can improve by a unilateral change of strategy. Here, of course, $V = x^*Ay^{*t}$, where A is the payoff matrix for Player I, so each such game has a definite *value*.

Notes

Uniqueness of equilibria. For the mixed-strategy extensions of the Matching Pennies and Rock–Paper–Scissors games, the arguments that proved the uniqueness of their Nash equilibria (Propositions 5.5 and 5.6) were shown to me by Paul Bourdon. The same is true for the "contrapositive" argument in the proof of Nash's theorem.

Mathematicians go to dinner. This game, which comes from Erica Klarreich's article [61], is a very simple version of a situation that game theorists call "The Tragedy of the Commons," more serious instances of which concern resource depletion and environmental degradation; see [47] for the original paper on this. In [61] Klarreich gives a nice nontechnical description of game theory and Nash's influence on it.

The Minimax Theorem. The original paper is [89]. According to Kuhn and Tucker [65] this theorem was

> ... the source of a broad spectrum of technical results, ranging from his [von Neumann's] extensions of the Brouwer Fixed-Point Theorem, developed for its proof, to new and unexpected methods for combinatorial problems.

von Neumann's above-mentioned extension of the Brouwer Fixed-Point Theorem is actually a precursor to a result of Kakutani's that Nash used to give a one-page proof of his existence theorem [85]. We'll study these matters in the next chapter.

The Nobel Prize. For his work on equilibrium in game theory Nash won the 1994 Nobel Prize in Economics. Quoting economist Roger Myerson: [81, esp. Sects. 1 and 6]:

> ... Nash's theory of noncooperative games should now be recognized as one of the outstanding intellectual advances of the twentieth century. The formulation of Nash equilibrium has had a fundamental and pervasive impact in economics and the social sciences which is comparable to that of the discovery of the DNA double helix in the biological sciences.

Nash's statement and proof (the one presented here) of his famous theorem occurred in his Ph.D. thesis [86, 87]. However the first published proof of this result was the short proof mentioned above, which appeared in the Proceedings of the National Academy of Sciences [85]. In an interesting article [79] written on the occasion of Nash's Nobel award, John Milnor surveys Nash's important contributions, not only to game theory, but also to algebraic and differential geometry, and to partial differential equations. Milnor also mentions the "Going to Dinner" game.

"That's trivial, you know. That's just a fixed-point theorem." Allegedly [84, p. 94] von Neumann's reply to Nash upon being shown the existence of what has since become known as Nash Equilibrium.

Schizophrenia. In his early thirties, at the height of his career, Nash succumbed to this dreaded disease and disappeared from scientific life for the next 30 or so years. Miraculously, he recovered in time to receive his 1994 Nobel Prize. About a month after the award ceremony, Nash's recovery was announced to the world by Sylvia Nasar in a *New York Times* article [83]. Nasar went on to write a full length biography of Nash [84], later adapted (very freely) into a major motion picture.

Chapter 6
Nash's "One-Page Proof"

KAKUTANI'S SET-VALUED FIXED-POINT THEOREM

Overview. In this chapter we'll study Shizuo Kakutani's extension of the Brouwer Fixed-Point Theorem to maps whose values are *sets*, and we'll show how John Nash used Kakutani's result to provide a very quick proof of his famous Theorem 5.11 on the existence of Nash Equilibrium.

Prerequisites. Undergraduate-level real analysis in $\mathbb{R}^N$, basic notions of set theory. The notions of normed linear spaces and metric spaces pop up, but nothing will be lost by thinking of everything in the context of $\mathbb{R}^N$.

6.1 Multifunctions

A *multifunction*[1] Φ from a set X to a set Y (notation: $\Phi: X \rightrightarrows Y$) is a function Φ defined on X whose values are *nonempty subsets* of Y.

We'll quite naturally refer to X as the *domain* of Φ, but perhaps somewhat less naturally will refer to Y as its *range*, and will further abuse terminology by defining $\Phi(E)$, the *image* of $E \subset X$, to be the *union* of the sets $\Phi(x)$ as x ranges through E. Proceeding in this vein, the *graph* of Φ will be the subset of the product set $X \times Y$ defined by

$$\text{graph}(\Phi) = \bigcup_{x \in X} \{x\} \times \Phi(x) = \{(x,y) \in X \times Y : y \in \Phi(x)\}.$$

Example 6.1 (Inverse Maps). Perhaps the most commonly encountered set-valued map is the inverse map induced by an ordinary function. For such a function $f: X \to Y$, define $f^{-1}: Y \rightrightarrows X$ by

$$f^{-1}(y) = \{x \in X : f(x) = y\} \qquad (y \in Y).$$

---

[1] Also called a "correspondence," or simply a "set-valued map."

© Springer International Publishing Switzerland 2016
J.H. Shapiro, *A Fixed-Point Farrago*, Universitext,
DOI 10.1007/978-3-319-27978-7_6

For example, if $X = \mathbb{R}$ and $f(x) = x^2$ for each $x \in \mathbb{R}$, then $f^{-1}(y) = \{\sqrt{y}, -\sqrt{y}\}$, a set-valued map whose domain is here taken to be the non-negative real numbers, and whose range is $\mathbb{R}$.

Exercise 6.1. For the general inverse mapping $f^{-1} \colon Y \rightrightarrows X$, what is graph (f^{-1}), and how does it relate to the graph of f? How does this play out for the particular map f^{-1} where $f(x) = x^2$ (x real)?

Exercise 6.2 (A step map). Consider the set-valued map $\Phi \colon \mathbb{R} \rightrightarrows \mathbb{R}$ defined by

$$\Phi(x) = \begin{cases} \{-1\} & \text{if } x < 0 \\ [-1, 1] & \text{if } x = 0 \\ \{+1\} & \text{if } x > 0 \end{cases}$$

What is $\Phi(\mathbb{R})$? Sketch the graph of Φ; how does it differ from that of the "usual" step function?

If $\Phi \colon X \rightrightarrows Y$ where X and Y are metric spaces, we'll say Φ is *closed-valued* if $\Phi(x)$ is closed in Y for every $x \in X$, with similar definitions applying to all other topological or geometric properties (openness, compactness, convexity, . . .) that sets may have. We'll say Φ has *closed graph* if graph (Φ) is closed in the product space $X \times Y$. For example, the step map defined in Exercise 6.2 above has closed graph.

Exercise 6.3. Show that if X and Y are metric spaces and $f \colon X \to Y$ is continuous, then the graph of $f^{-1} \colon Y \rightrightarrows X$ is closed in $Y \times X$.

Exercise 6.4. Suppose X and Y are metric spaces. Show that if $\Phi \colon X \rightrightarrows Y$ has closed graph then $\Phi(x)$ is closed for each $x \in X$. Is the converse true?

Example 6.2 (The metric projection). Suppose (X, d) is a metric space and K is a compact subset of X. Then for each $x \in X$ there is an element $\xi \in K$ that is nearest to x, i.e.,

$$\inf_{k \in K} d(x, k) = d(x, \xi).$$

The *metric projection* of X onto K is the set-valued map P_K that associates to a point of X the collection of all these elements of K that are nearest to x.

We've already encountered the metric projection in Sect. 4.4, where $X = \mathbb{R}^N$ and K is both compact and convex. In that setting we called P_K the "closest point map," and showed that it's an "ordinary" function in that each of its values is a singleton.

Exercise 6.5. Prove that the metric projection of a metric space onto a compact subset has closed graph.

6.2 Best Response

The notion of "best response" arose during our work in Chap. 5 on Nash Equilibrium. In this section we'll strip the concept down to its essentials and see how it really concerns a multifunction. We begin with an $M \times N$ real matrix A, and its associated quadratic function

$$u(x,y) = xAy^t \qquad (x \in \mathbb{R}^M, y \in \mathbb{R}^N).$$

Exercise 6.6. Show that u is continuous on $\mathbb{R}^M \times \mathbb{R}^N$.

We'll be concerned only with the values u takes on the cartesian product $\Pi_M \times \Pi_N$ of the standard simplices in $\mathbb{R}^M$ and $\mathbb{R}^N$, respectively, where—as in the last chapter— we think of each simplex as a collection of probability vectors representing mixed strategies for a two-player game, and $u(x,y)$ as Player I's "payoff" for playing strategy $x \in \Pi_M$ against Player II's strategy $y \in \Pi_N$. Recall that for each $y \in \Pi_N$ the function $u(\cdot,y)$ is continuous on the compact subset Π_M of $\mathbb{R}^M$, hence there exists $x^* \in \Pi_M$ such that $u(x^*,y) = \max_{x \in \Pi_m} u(x,y)$.

We've previously called the probability vector x^* a "best response" for the first player to the second player's strategy y. In this section we'll consider the set $\mathrm{BR}(y)$ of *all* best responses to y, thereby obtaining the "best-response multifunction" $\mathrm{BR} \colon \Pi_N \rightrightarrows \Pi_M$ for the payoff-matrix A.

Proposition 6.3. *Each value of BR is a nonvoid compact, convex subset of Π_M. The graph of BR is closed in $\Pi_N \times \Pi_M$.*

Proof. We've already noted that $\mathrm{BR}(y)$ is nonempty for each $y \in \Pi_N$. Let's fix y and write $u_y(\cdot) = u(\cdot,y)$, so u_y is an affine continuous function on Π_M. Let μ be the first player's payoff for some best response (hence for *all* best responses) to the second player's strategy y, i.e., $\mu = u(x^*,y) = u_y(x^*)$ for every $x^* \in \mathrm{BR}(y)$. Thus $\mathrm{BR}(y) = u_y^{-1}(\{\mu\})$. Now for each continuous function the inverse image of a closed set is closed, and for each affine map the inverse image of a convex set is convex (Exercise 5.2, page 59). Thus $\mathrm{BR}(y)$, being the image under u_y^{-1} of the closed, convex singleton $\{\mu\}$, is itself a convex subset of Π_M that is closed, and hence compact, in Π_M.

As for the closed graph, suppose $((y_j,x_j))_{j=1}^\infty$ is a sequence of points in the graph of BR that converges in $\Pi_N \times \Pi_M$ to a point (y,x), necessarily also in $\Pi_N \times \Pi_M$. We need to prove that (y,x) belongs to the graph of B, i.e., that x is a best response to y. For each j we know that x_j is a best response to y_j, i.e., if $\xi \in \Pi_M$ is fixed, then $u(x_j,y_j) \geq u(\xi,y_j)$. By the continuity of u (Exercise 6.6) this inequality is preserved as $j \to \infty$, so $u(x,y) \geq u(\xi,y)$. Since this last inequality is true for each $\xi \in \Pi_M$, the strategy x is a best response to y, i.e., (y,x) belongs to the graph of BR. $\qquad\square$

6.3 The Kakutani Fixed-Point Theorem

To say $x \in X$ is a *fixed point* of $\Phi \colon X \rightrightarrows X$ means that $x \in \Phi(x)$.

Exercise 6.7 (Fixed points of some multifunctions).

(a) For an "ordinary" function $f \colon X \to X$, what does it mean to say that $x \in X$ is a fixed point of $f^{-1} \colon X \rightrightarrows X$?

(b) What are the fixed points of the step function of Exercise 6.2?

(c) For the metric projection as in Example 6.2, what are the fixed points?

Here is the main theorem of this chapter.

Theorem 6.4 (Kakutani's set-valued fixed-point theorem). *If C is a nonvoid compact, convex subset of $\mathbb{R}^N$ and $\Phi \colon C \rightrightarrows C$ is a multifunction with compact, convex values and closed graph, then Φ has a fixed point.*

Were Φ to have a "continuous selection," i.e., a continuous map $\varphi \colon C \to C$ with $\varphi(x) \in \Phi(x)$ for each $x \in C$, then Kakutani's result would follow immediately from the Brouwer Fixed-Point Theorem. However under the hypotheses of Kakutani's theorem such selections need not always exist. For example, the "step-map" Φ of Exercise 6.2, when restricted to (say) the closed interval $C = [-1,1]$, satisfies those hypotheses, but has no continuous selection. Nevertheless there is an "approximate" version of this argument that will carry the day.

To simplify the statement of this "approximate-selection theorem:" for (X,d) a metric space, $E \subset X$, and $\varepsilon > 0$, define $[E]_\varepsilon = \bigcup_{x \in E} B_\varepsilon(x)$, where $B_\varepsilon(x)$ is the open d-ball in X of radius ε, centered at x. Thus $[E]_\varepsilon$ is the set of all points in X that lie a distance $< \varepsilon$ from E. We'll call $[E]_\varepsilon$ the "ε-expansion of E"; clearly, it's an open subset of X.

Exercise 6.8. Suppose C is a convex subset of $\mathbb{R}^N$ (or for that matter, of a normed linear space). Show that the ε-expansion $[C]_\varepsilon$ is convex for every $\varepsilon > 0$.

When the space in which such expansions take place is the product of two metric spaces, we'll take the "product distance" to be the maximum of the distances in the individual factors. More precisely: if (X,ρ) and (Y,σ) are metric spaces, then the metric in $X \times Y$ will be

$$d((x,y),(x',y')) \colon = \max\left[\rho(x,x'),\sigma(y,y')\right] \qquad ((x,y),(x',y') \in X \times Y).$$

Exercise 6.9. For the product space $\mathbb{R} \times \mathbb{R}$ (with the usual metric on each factor) describe and sketch the open ε-ball with center at the origin. What's the relationship between this "product metric" and the Euclidean metric? Do they have the same convergent sequences?

Theorem 6.5 (The Approximate-Selection Theorem). *Suppose X is a compact metric space, Y a nonvoid compact, convex subset of a normed linear space, and $\Phi \colon X \rightrightarrows Y$ a multifunction all of whose values are compact and convex. If $\mathrm{graph}(\Phi)$ is closed, then for every $\varepsilon > 0$ there exists a continuous function $f \colon X \to Y$ with $\mathrm{graph}(f) \subset [\mathrm{graph}(\Phi)]_\varepsilon$.*

Proof of Kakutani's Theorem. Granting the truth of The Approximate-Selection Theorem, use that theorem to provide, for each positive integer n, a continuous function $f_n \colon C \to C$ whose graph lies in $[\mathrm{graph}\,(\Phi)]_{1/n}$. Brouwer's theorem provides, for each n, a fixed point p_n for f_n, hence the pair (p_n, p_n) lies in graph (f_n), and so lies within $1/n$ of graph (Φ). Since C is compact the sequence of "approximate fixed points" (p_n) has a convergent subsequence—say convergent to $p \in C$. But then the corresponding subsequence of pairs (p_n, p_n) converges in $C \times C$ to (p, p), which lies a distance zero from graph (Φ). Since graph (Φ) is closed, (p, p) belongs to it, i.e., $p \in \Phi(p)$. $\qquad\square$

For the proof of the Approximate-Selection Theorem we need to explore the consequences for a multifunction of possessing a closed graph.

Lemma 6.6 (Semicontinuity Lemma). *Suppose X and Y are metric spaces with Y compact. If $\Phi \colon X \rightrightarrows Y$ has closed graph, then for every $\varepsilon > 0$ and $x \in X$ there exists a ball B_x centered at x such that $\Phi(B_x) \subset [\Phi(x)]_\varepsilon$.*

Multifunctions satisfying the conclusion of this Proposition are said to be *upper semicontinuous*, reflecting the fact that for ξ close to x the set $\Phi(\xi)$ "cannot be too large" compared to $\Phi(x)$.

Proof of Lemma. Suppose Φ is *not* upper semicontinuous. Then there exists $\varepsilon > 0$, $x \in X$, and a sequence (x_n) in X convergent to x such that $\Phi(x_n)$ is not contained in $[\Phi(x)]_\varepsilon$, i.e., for each index n there exists $y_n \in \Phi(x_n)$ that lies *at least* ε-distant from $\Phi(x)$. Since Y is compact, we may, upon passing to an appropriate subsequence, assume that (y_n) converges to some point $y \in Y$, which also lies at least ε-distant from $\Phi(x)$. Thus the sequence of pairs (x_n, y_n), which belongs to graph (Φ), converges in $X \times Y$ to the pair (x, y) which is *not* in graph (Φ), so graph (Φ) is not closed. $\qquad\square$

Exercise 6.10 (Converse to the Semicontinuity Lemma). Suppose X and Y are metric spaces and that $\Phi \colon X \rightrightarrows Y$ is closed-valued and upper semicontinuous. Show that graph (Φ) is closed.

Proof of Theorem 6.5. Fix $\varepsilon > 0$. For each $x \in X$, Lemma 6.6 provides us with a ball B_x centered at x for which $\Phi(B_x) \subset [\Phi(x)]_\varepsilon$. We may take this ball to have radius $< \varepsilon$. The collection of balls $\frac{1}{2}B_x$ (center still at x, but radius half that of B_x) is an open cover of the compact space X, and so has a finite subcover which, to save on subscripts, we'll write as $\mathscr{B} = \{\frac{1}{2}B_1, \frac{1}{2}B_2, \dots \frac{1}{2}B_n\}$, where B_j is the original ball centered at $x_j \in X$. By Proposition B.6 of Appendix B there is a partition of unity $\{p_1, p_2, \dots p_n\}$ subordinate to the covering $\mathscr{B}$, i.e., for each index j the function $p_j \colon X \to [0, 1]$ is continuous and vanishes off $\frac{1}{2}B_j$, while the totality of these functions sums to 1 on X. For each index j choose $y_j \in \Phi(x_j)$; then define $f \colon X \to Y$ by

$$f(x) = \sum_{j=1}^n p_j(x) y_j \qquad (x \in X). \tag{6.1}$$

Thus f is continuous on X, with values in the convex hull of $\Phi(X)$. We'll be done if we can show that graph $(f) \subset [\mathrm{graph}\,(\Phi)]_\varepsilon$, i.e., that for $x \in X$ the point $(x, f(x))$ of graph (f) lies within ε of graph (Φ).

Fix $x \in X$ and note that since the "partition function" p_j vanishes identically off $\frac{1}{2}B_j$, the sum on the right-hand side of Eq. (6.1) involves only those indices j for which $x \in \frac{1}{2}B_j$. Let $J = J(x)$ denote this collection of indices, so that $f(x) = \sum_{j \in J} p_j(x) y_j$.

Let m be an index in $J(x)$ corresponding to a ball B_j of largest radius for $j \in J(x)$. Note that for each $j \in J(x)$ our point x lies within radius$(\frac{1}{2}B_j) \leq$ radius$(\frac{1}{2}B_m)$ of the center x_j, so all these points x_j for $j \in J(x)$ lie within radius(B_m) of each other. Thus for each $j \in J(x)$ we have (from our semicontinuity-driven initial choice of balls): $\Phi(x_j) \subset [\Phi(x_m)]_\varepsilon$. In particular the point y_j, chosen to lie in $\Phi(x_j)$, also lies in $[\Phi(x_m)]_\varepsilon$. *Conclusion:* $f(x)$ lies in the convex hull of $[\Phi(x_m)]_\varepsilon$. But $\Phi(x_m)$ is convex, hence so is its ε-expansion $[\Phi(x_m)]_\varepsilon$ (Exercise 6.8). Thus $f(x) \in [\Phi(x_m)]_\varepsilon$, i.e., there exists $y \in \Phi(x_m)$ with $d(f(x), y) < \varepsilon$.

Finally, note that x, being a point of $\frac{1}{2}B_m$, lies within radius$(B_m) < \varepsilon$ of x_m, so by our definition of the metric in $X \times Y$ the point $(x, f(x))$ lies within ε of $(x_m, y) \in$ graph(Φ), i.e., $(x, f(x))$ lies within ε of graph(Φ), as promised.   □

Remark. In this proof we used the compactness of Y only to deduce the upper semicontinuity of Φ from the closed-graph hypothesis. Thus we could eliminate this extra assumption on Y by requiring at the outset that Φ be upper semicontinuous.

6.4 Application to Nash Equilibrium

We're now in a position to give Nash's "one-page" proof [85] of Theorem 5.11 on the existence of Nash Equilibria for mixed-strategy extensions of non-cooperative finite games.

As usual, we'll keep notation to a minimum by concentrating on the two-person situation. Here the original "pure strategy" game provides M strategies for Player I and N strategies for Player II. The payoff matrices are A and B for the respective players, and the mixed-strategy payoff functions are: $u_{\mathrm{I}}(x, y) = xAy^t$ for Player I, and $u_{\mathrm{II}}(x, y) = xBy^t$, where $x \in \Pi_M$ and $y \in \Pi_N$. Denote the best response multifunction for Player I by BR$_{\mathrm{I}}$, and for Player II by BR$_{\mathrm{II}}$. Recall that to say the strategy pair $(x^*, y^*) \in \Pi_M \times \Pi_N$ is a *Nash Equilibrium* means that each is a best response to the other, i.e., that $x^* \in$ BR$_{\mathrm{I}}(y^*)$ and $y^* \in$ BR$_{\mathrm{II}}(x^*)$. So if we define BR$: \Pi_M \times \Pi_N \rightrightarrows \Pi_M \times \Pi_N$ by

$$\mathrm{BR}(x, y) = \mathrm{BR}_{\mathrm{I}}(y) \times \mathrm{BR}_{\mathrm{II}}(x) \qquad (x, y) \in \Pi_M \times \Pi_N,$$

then we're saying that our strategy pair (x^*, y^*) is a Nash Equilibrium if and only if it's a fixed point of the multifunction *BR*.

Since the set $\Pi_M \times \Pi_N$ on which *BR* is acting is a compact, convex subset of $\mathbb{R}^{MN}$, Kakutani's Theorem 6.4 will produce the desired fixed point if we can verify that *BR* satisfies its hypotheses.

To this end, note that we've already shown (Proposition 6.3) the individual cartesian factors BR_I and BR_{II} of BR to have nonvoid compact convex values. Since these properties are preserved by cartesian products the values of BR are also nonvoid, compact, and convex. We've also shown that the factors of BR each have graphs that are closed in their ambient compact product spaces, hence they are compact. Now the graph of BR is homeomorphic via a permutation of coordinates to the cartesian product of the graphs of BR_I and BR_{II}, so it too is compact. Thus Kakutani's theorem applies, and shows that BR has the desired fixed point. □

Notes

Kakutani's Theorem. Theorem 6.5, the approximate selection result that did the heavy lifting in the proof of Kakutani's Theorem (Theorem 6.4) is attributed variously to von Neumann [89] and more recently to Cellina [23]. The proof given here is taken from [16, pp. 59–62].

Kakutani's original proof [58]. This takes place on an N-simplex where Kakutani constructs, for each triangulation, a piecewise-affine approximate selection.

Another famous Kakutani. Michiko Kakutani, the influential New York Times literary critic and 1998 Pulitzer Prize winner, is Shizuo Kakutani's daughter.

von Neumann and Kakutani. As mentioned in the *Notes* to the previous chapter, von Neumann proved a special case of Theorem 6.4 on which he based his proof of the Minimax Theorem. The reference for this is [89].

Part III
Beyond Brouwer: Dimension $= \infty$

The setting now shifts to infinite dimensional normed linear spaces. Here we'll prove Schauder's extension of the Brouwer Fixed-Point Theorem and will explore some of its applications to initial-value problems, operator theory, and measure theory.

Chapter 7
The Schauder Fixed-Point Theorem

An Infinite Dimensional Brouwer Theorem

Overview. Recall that to say a metric space has the *fixed-point property* means that every continuous mapping taking the space into itself must have a fixed point. In Chap. 4 we proved two versions of the Brouwer Fixed-Point Theorem:

> The "Ball" version (Theorem 4.1). *The closed unit ball of $\mathbb{R}^N$ has the fixed-point property,*

and the seemingly more general, but in fact equivalent

> "Convex" version (Theorem 4.5). *Every compact convex subset of $\mathbb{R}^N$ has the fixed-point property.*

It turns out that the "ball" version of Brouwer's theorem does not survive the transition to infinitely many dimensions. However all is not lost: the "convex" version *does* survive: compact, convex subsets of normed linear space *do* have the fixed-point property. This is the famous Schauder Fixed-Point Theorem (circa 1930) which will occupy us throughout this chapter. After proving the theorem we'll use it to prove an important generalization of the Picard–Lindelöf Theorem of Chap. 3 (Theorem 3.10). The Schauder Theorem will also be important in the next chapter where it will provide a key step in the proof of Lomonosov's famous theorem on invariant subspaces for linear operators on Banach spaces.

Prerequisites. Basics of normed linear spaces and compactness in metric spaces.

7.1 The Theorem

Theorem 7.1 (The Schauder Fixed-Point Theorem)**.** *In every normed linear space, each compact, convex subset has the fixed-point property.*

© Springer International Publishing Switzerland 2016
J.H. Shapiro, *A Fixed-Point Farrago*, Universitext,
DOI 10.1007/978-3-319-27978-7_7

Our proof of Brouwer's theorem depended strongly on the compactness of the closed unit ball of $\mathbb{R}^N$. We'll see in the next chapter (Proposition 8.7) that *no infinite dimensional normed space has this property*. This is easiest to understand for infinite dimensional Hilbert space, where for each orthonormal sequence the distance between two distinct elements is $\sqrt{2}$, hence such a sequence, which belongs to the closed unit ball, has no convergent subsequence. Conclusion: The closed unit ball of an infinite dimensional Hilbert space is non-compact.

It gets worse! Recall that every separable Hilbert space is isometrically isomorphic to ℓ^2, the Hilbert space consisting of square-summable scalar sequences endowed with the norm

$$\|f\| = \left(\sum_{n=1}^{\infty} |f(n)|^2 \right)^{1/2} \qquad (f = (f(n))_1^{\infty} \in \ell^2).$$

Proposition 7.2. *The closed unit ball of the Hilbert space ℓ^2 (hence of every separable Hilbert space) does not have the fixed-point property.*

Proof. Let $(e_n)_1^{\infty}$ be the standard basis in ℓ^2 (i.e., e_n is the sequence with 1 in the n-th position and zero elsewhere), and define the map T on ℓ^2 by

$$Tf = (1 - \|f\|)e_1 + \sum_{n=1}^{\infty} f(n)e_{n+1} = (1 - \|f\|, f(1), f(2), \ldots) \qquad (f \in \ell^2).$$

Then for $f \in \ell^2$ with $\|f\| \leq 1$ we have from the triangle inequality:

$$\|Tf\| \leq (1 - \|f\|) + \underbrace{\left\| \sum_{n=1}^{\infty} f(n)e_{n+1} \right\|}_{=\|f\|} = (1 - \|f\|) + \|f\| = 1,$$

so T takes the closed unit ball B of ℓ^2 into itself. Furthermore, if f and g are two vectors in ℓ^2, then a straightforward calculation shows

$$\|Tf - Tg\|^2 = (\|f\| - \|g\|)^2 + \|f - g\|^2 \tag{7.1}$$

from which one deduces that T is continuous on ℓ^2.

Claim. *T has no fixed point in B.*

Proof of Claim. Suppose $f \in B$ were a fixed point of T. Upon equating components in the equation $Tf = f$ we would obtain

$$f(n) = 1 - \|f\| \qquad (n \in \mathbb{N}), \tag{7.2}$$

thus exhibiting f as a constant function. But $f \in \ell^2$, so $f(n) \to 0$, hence $f(n) = 0$ for all n. This, along with Eq. (7.2) above, yields the contradiction $0 = 1$. $\qquad\square$

Exercise 7.1. Verify the "straightforward calculation" (7.1) and show that it really does establish the continuity claimed above for T.

The key to proving the Schauder Fixed-Point Theorem will be to show that each compact subset of a normed linear space can be "almost" embedded in the convex hull of a finite subset of its points. Such convex hulls are compact (Proposition C.5 of Appendix C), so Brouwer's theorem can be applied to produce approximate fixed points, and hence by Lemma 2.2 (the "Approximate Fixed-Point Lemma," p. 24), an actual fixed point.

7.2 The Proof

Crucial to the proof of Schauder's theorem is the easily proved observation that each compact subset of a metric space can be approximated arbitrarily closely by a finite set. More precisely, for every $\varepsilon > 0$ our compact set K contains an ε-*net*: a finite subset F of K such that for each point $x \in K$ there is a point of F lying within ε of x (Proposition B.3 of Appendix B). This in turn gives rise to an important map called the *Schauder Projection*.

Proposition 7.3 (The Schauder Projection). *Suppose C is a compact convex subset of a normed linear space X. Then given $\varepsilon > 0$ and an ε-net F_ε contained in C, there exists a continuous map P_ε (the* Schauder Projection*) that takes C into the convex hull of F_ε such that $\|P_\varepsilon(x) - x\| < \varepsilon$ for every $x \in C$.*

Proof. We are assuming that $F_\varepsilon = \{x_1, x_2, \ldots x_N\} \subset C \subset \bigcup_{j=1}^{N} B(x_j, \varepsilon)$, where $B(x_j, \varepsilon)$ denotes the open ball of radius ε centered at the point x_j. By Proposition B.6 (p. 190) there is a partition of unity $(p_1, p_2, \ldots p_N)$ on C subordinate to the covering $\{B(x_j, \varepsilon) : 1 \leq j \leq N\}$. Specifically: $\sum_j p_j \equiv 1$ on C and for each index j the function p_j is non-negative and continuous on C, and identically zero outside $B(x_j, \varepsilon)$.

Now we proceed as in the proof of the "Approximate Selection Theorem" of the last chapter (Theorem 6.5): define the map P_ε on C by

$$P_\varepsilon(x) = \sum_{j=1}^{N} p_j(x)x_j \qquad (x \in C).$$

For each $x \in C$ the vector $P_\varepsilon(x)$ is a convex combination of the vectors x_j, hence P_ε maps C into the convex hull of F_ε (Proposition C.4 of Appendix C), and since each coefficient function p_j is continuous on C so is P_ε. Moreover

$$\|P_\varepsilon(x) - x\| = \left\| \sum_{j=1}^{N} p_j(x)(x_j - x) \right\| \leq \sum_{j=1}^{N} p_j(x)\|x_j - x\| \qquad (x \in C),$$

where in the last sum on the right the coefficient $p_j(x)$ is zero whenever $\|x - x_j\|$ is $\geq \varepsilon$. Thus $\|P_\varepsilon(x) - x\| < \varepsilon \sum_j p_j(x) = \varepsilon$ for every $x \in C$, as desired. $\qquad \square$

The final estimate above could also be viewed like this: The coefficient $p_j(x)$ vanishes for those vectors $x_j - x$ that lie outside the ball $B(0,\varepsilon)$, so $P_\varepsilon(x) - x$ is a "subconvex" combination of points in that ball, so also lies in that ball.

Proof of the Schauder Fixed-Point Theorem. We're given a compact, convex subset C of a normed space X and a continuous map $f \colon C \to C$. We wish to show that f has a fixed point. By the Approximate Fixed-Point Lemma (Lemma 2.2, p. 24) it's enough to show that: given $\varepsilon > 0$ there exists $x_\varepsilon \in C$ such that $\|f(x_\varepsilon) - x_\varepsilon\| < \varepsilon$.

To this end let $\varepsilon > 0$ be given, choose an ε-net $F_\varepsilon \subset C$, and let P_ε be the Schauder projection of C onto $\operatorname{conv}(F_\varepsilon)$. Then $g_\varepsilon = P_\varepsilon \circ f$ maps C continuously into $\operatorname{conv}(F_\varepsilon)$, and so maps $\operatorname{conv}(F_\varepsilon)$ continuously into itself. Since $\operatorname{conv}(F_\varepsilon)$ is a compact (Proposition C.5 of Appendix C), convex subset of a finite dimensional subspace of X, it is, by Proposition C.9 (p. 197) homeomorphic (even linearly) to a compact, convex subset of a finite dimensional Euclidean space, so by the "Convex" Brouwer Fixed-Point Theorem (Theorem 4.5), g_ε has a fixed point x_ε that lies in $\operatorname{conv}(F_\varepsilon)$, and hence in C. Thus:

$$\|f(x_\varepsilon) - x_\varepsilon\| = \|f(x_\varepsilon) - g_\varepsilon(x_\varepsilon)\| = \|f(x_\varepsilon) - P_\varepsilon(f(x_\varepsilon))\| < \varepsilon,$$

as desired.   □

7.3 Generalization to Non-compact Situations

Here's a sobering thought about the Schauder Fixed-Point Theorem: In infinite dimensional normed linear spaces there are not many compact sets. For example, we noted just after our statement of the Schauder Theorem (Theorem 7.1, p. 75) that in such spaces closed balls are never compact, and we gave an argument to prove this for Hilbert space. The exercise below asks you to prove this for the situation we'll encounter in the next section.

> *Exercise* 7.2. Suppose I is a compact interval of the real line. Show that no closed ball in $C(I)$ is compact.

All is not lost, however: thanks to the following result, Schauder's theorem can be applied to non-compact situations—provided that the maps in question have some "built-in compactness." Recall that to say a set in a metric space is "relatively compact" means that its closure is compact.

Corollary 7.4. *Suppose C is a closed convex subset of Banach space and $f \colon C \to C$ is a continuous map. If $f(C)$ is relatively compact in C then f has a fixed point.*

Proof. Since $f(C)$ is a relatively compact subset of the convex set C, Proposition C.6 of Appendix C guarantees that its convex hull is relatively compact. Thus the closure, K, of $\operatorname{conv}(f(C))$ is compact in our Banach space, and since C is closed, $K \subset C$. Thus $f(K) \subset f(C) \subset K$, so Schauder's Theorem applies to the restriction of f to K, and produces the desired fixed point.   □

7.4 Application: Initial Value Problems

In Sect. 3.4 we used the Banach Contraction-Mapping Principle to prove that each initial-value problem of the form

$$y' = f(x,y), \quad y(x_0) = y_0, \qquad \text{(IVP)}$$

with f satisfying appropriate smoothness conditions, has a unique solution on some nontrivial interval centered at x_0. The conditions required by this " Picard–Lindelöf theorem" (Theorem 3.10, p. 35) were that f be defined and continuous on some open subset U of $\mathbb{R}^2$ containing the point (x_0, y_0), and that $\frac{\partial f}{\partial y}$, the partial derivative of f with respect to the second variable, exist on U and be continuous there (or just that f satisfy a uniform Lipschitz condition on U in the second variable). Now, thanks to the Schauder Fixed-Point Theorem, we'll be able to prove the existence of solutions to (IVP) *without having to assume extra second-variable smoothness for f*. However there will be a cost: the solutions need no longer be unique!

Theorem 7.5 (Peano's Theorem). *Suppose f is a real-valued function that is continuous on some open subset of $\mathbb{R}^2$ containing the point (x_0, y_0). Then the initial-value problem* (IVP) *has a solution on some nontrivial interval centered at x_0.*

Proof. By the work of Sect. 3.4 through Lemma 3.9 (p. 34) we know that there is a compact real interval I centered at x_0 such that the Banach space $C(I)$ contains a closed ball $\overline{B}$ centered at the constant function y_0 with $T(\overline{B}) \subset \overline{B}$, where T is the continuous integral operator on $C(I)$ defined by

$$Tu(x) = y_0 + \int_{t=x_0}^{x} f(t, u(t))\, dt \qquad (u \in C(I)).$$

We saw in Sect. 1.3 (p. 4) that a function $y = u(x)$ is a solution on I of (IVP) if and only if it is a fixed point of T. The Schauder Fixed-Point Theorem would immediately provide such a fixed point if only $\overline{B}$ were compact in $C(I)$. Unfortunately Exercise 7.2 above shows that it's not! However, we'll be able to show that $T(\overline{B})$ is relatively compact in $C(I)$, so the existence of a fixed point for T, and therefore of a solution for (IVP), will follow from Corollary 7.4.

   To prove this relative compactness it's enough to show, by the Arzela–Ascoli Theorem (Theorem B.8 of Appendix B), that $T(\overline{B})$ is bounded in $C(I)$ and *equicontinuous* on I, i.e., for every $\varepsilon > 0$ there exists $\delta > 0$ such that

$$x, y \in I \ \& \ |x - y| < \delta \implies |Tu(x) - Tu(y)| < \varepsilon \qquad \forall u \in \overline{B}.$$

The boundedness of $T(\overline{B})$ has already been established, since $T(\overline{B}) \subset \overline{B}$. As for equicontinuity: fix $u \in \overline{B}$ and note that for each pair of points $x, y \in I$ with $x \leq y$ and $|x - y| < \delta$:

$$|Tu(x) - Tu(y)| = \left| \int_{t=x_0}^{x} f(t, u(t)) \, dt - \int_{t=x_0}^{y} f(t, u(t)) \, dt \right|$$

$$= \left| \int_{t=x}^{y} f(t, u(t)) \, dt \right|$$

$$\leq \int_{t=x}^{y} |f(t, u(t))| \, dt$$

$$\leq M|x - y| \, .$$

Thus if $\varepsilon > 0$ is given and $|x - y| < \varepsilon/M$, then $|Tu(x) - Tu(y)| < \varepsilon$ for every $u \in \overline{B}$, thus establishing the equicontinuity of $T(\overline{B})$.

Now the equicontinuity of $T(\overline{B})$ carries immediately over to its convex hull $\mathrm{conv}\,(T(\overline{B}))$, which, being contained in B, is also bounded. Thus $\mathrm{conv}\,(T(\overline{B}))$ is relatively compact, so K, its closure in $C(I)$, is compact. Since $T(\overline{B}) \subset K$, Schauder's theorem applies to the restriction of T to K, and furnishes the desired fixed point. $\quad\square$

Non-Uniqueness in Peano's Theorem. In contrast to the Banach Contraction-Mapping Principle, Schauder's Theorem makes no claims about uniqueness for the fixed point it produces. The example below shows that non-uniqueness can even occur "naturally." Consider the initial-value problem:

$$y' = -2y^{1/2}, \; y(0) = 1, \; t \geq 0. \tag{T}$$

The solution that comes immediately to mind is: $y(t) = (1 - t)^2$. Here's another one:

$$y(t) = \begin{cases} (1 - t)^2 & (0 \leq t \leq 1) \\ 0 & (t \geq 1). \end{cases}$$

The initial-value problem (T) expresses a physical phenomenon discovered by Evangalista Toricelli (1608–1647). *Toricelli's Law* states that water issues from a small hole in the bottom of a container at a rate that is proportional to the square root of the water's depth (see, e.g., [33] for more details). In (T) the function $y(t)$ expresses the depth of the water in the container at time $t \geq 0$. The second solution to (T) is the realistic one for this interpretation; it asserts that the water starts out at $t = 0$ with height 1 and flows out until the container runs dry at $t = 1$, and thereafter stays dry. By contrast, the "obvious" solution $y = (1 - t)^2$ for all $t \geq 0$ unrealistically predicts that after the tank runs dry at $t = 1$ it miraculously starts filling up again.

7.5 Application: Multifunctions Again

In the proof of the Kakutani Fixed-Point Theorem (Theorem 6.4, p. 68): if one replaces $\mathbb{R}^N$ with a normed linear space, and Brouwer's Fixed-Point Theorem with Schauder's, then the argument goes through without further change, yielding the following generalization:

Theorem 7.6 (A "Kakutani–Schauder" fixed-point theorem). *If C is a nonvoid compact, convex subset of a normed linear space and $\Phi : C \rightrightarrows C$ is a multifunction with compact, convex values, and closed graph, then Φ has a fixed point.*

Notes

The Schauder Fixed-Point Theorem. Schauder published this result in [107, 1930].

Generalizations of Schauder's theorem. In 1935 Andrey Tychonoff [120] generalized Schauder's Theorem to arbitrary linear topological spaces (see Sect. 9.3, p. 106 for the definition) that are *locally convex,* i.e., for which the neighborhoods of each point have a basis of convex sets. That same year Schauder posed the problem of extending his theorem to complete, metrizable, linear topological spaces. This "Schauder Conjecture" remained open until 2001 when it was settled in the affirmative by Robert Cauty [22]. In [31] Tadeusz Dobrowolski offers an expanded exposition of Cauty's work, along with further references and historical background.

Failure of the fixed-point property for non-compact convex sets. Proposition 7.2, showing that the closed unit ball of infinite dimensional Hilbert space fails to have the fixed-point property, is due to Kakutani[59, 1943]. The generalization to all infinite dimensional normed linear spaces was proved in 1951 by Dugundji [34, Theorem 6.3, p. 362], who showed that in this setting the closed unit ball can always be retracted onto its boundary. Later Victor Klee generalized Dugundji's result even further [62, Sect. 2.3] by showing that *for every metrizable locally convex linear topological space: if a convex set has the fixed-point property, then it must be compact.* Along with Tychonoff's extension of the Schauder theorem, this characterizes for metrizable locally convex spaces the convex sets with the fixed-point property; they are precisely the compact ones.

Chapter 8
The Invariant Subspace Problem

LOMONOSOV'S FAMOUS THEOREM

Overview. This chapter is about the most vexing problem in the theory of linear operators on Hilbert space:

THE INVARIANT SUBSPACE PROBLEM. *Does every operator on Hilbert space have a nontrivial invariant subspace?*

Here "operator" means "continuous linear transformation," and "invariant subspace" means "closed (linear) subspace that the operator takes into itself." To say that a subspace is "nontrivial" means that it is neither the zero subspace nor the whole space. Examples constructed toward the end of the last century show that in the generality of Banach spaces there do exist operators with only trivial invariant subspaces. For *Hilbert space*, however, the Invariant Subspace Problem remains open, and is the subject of much research. In this chapter we'll see why invariant subspaces are of interest and then will prove one of the subject's landmark theorems: Victor Lomonosov's 1973 result, a special case of which states:

If an operator T on a Banach space commutes with a non-zero compact operator, then T has a nontrivial invariant subspace.

This result, which far surpassed anything that seemed attainable at the time, is only part of what Lomonosov proved in an astonishing two-page paper [71] that introduced nonlinear methods—in particular the Schauder Fixed-Point Theorem—into this supposedly hard-core-linear area of mathematics.

Prerequisites. Basics of inner-product spaces, Hilbert and Banach spaces.

© Springer International Publishing Switzerland 2016
J.H. Shapiro, *A Fixed-Point Farrago*, Universitext,
DOI 10.1007/978-3-319-27978-7_8

8.1 Invariant Subspaces

For linear transformations on vector spaces for which no topology is assumed, "invariant subspace" will simply mean "subspace taken into itself by the transformation." Eigenvalues give rise to an important class of nontrivial invariant subspaces.

Theorem 8.1 (Invariance of eigenspaces). *Suppose T is a linear transformation on a vector space V, and that T is not a scalar multiple of the identity. Let λ be an eigenvalue of T. Then the subspace $\ker(T - \lambda I)$ is nontrivial and invariant for every linear transformation on V that commutes with T.*

Proof. Let $E = \ker T - \lambda I$. By hypothesis there is a vector $v \in V \backslash \{0\}$ with $Tv = \lambda v$. Thus $v \in E$, so $E \neq \{0\}$. Since $T \neq \lambda I$ we know $E \neq V$. Thus E is nontrivial.

Now suppose S is a linear transformation on V that commutes with T. Suppose $v \in E$. We wish to show that $Sv \in E$, i.e., that $TSv - \lambda Sv = 0$. This follows right away from the commutativity of S and T:

$$TSv - \lambda Sv = STv - \lambda Sv = S(\lambda v) - \lambda Sv = \lambda Sv - \lambda Sv = 0. \qquad \square$$

Exercise 8.1 (Invariant subspaces without eigenvalues). Let $C([0,1])$ denote the Banach space of complex-valued continuous functions on the unit interval $[0,1]$, endowed with the "max-norm"
$$\|f\| = \max\{|f(x)| : 0 \leq x \leq 1\} \qquad (f \in C([0,1])).$$
Show that the *Volterra operator*, defined by

$$V f(x) = \int_0^x f(t) \, dt \qquad (f \in C([0,1]), \; x \in [0,1]),$$

is an operator that takes the Banach space $C([0,1])$ into itself, that has no eigenvalue, but that nonetheless has nontrivial invariant subspaces.

Hyperinvariant subspaces. If a subspace of a Banach space is invariant for every operator that commutes with a given operator T, we'll call that subspace *hyperinvariant* for T. Thus Theorem 8.1 shows that every operator on $\mathbb{C}^N$ that's not a scalar multiple of the identity has a nontrivial hyperinvariant subspace. It's not known, however, if this is true for infinite dimensional Hilbert spaces. In other words, the "Hyperinvariant Subspace Problem" is just as open as is the "Invariant Subspace Problem."

Why the Invariant Subspace Problem? In studying the Invariant Subspace Problem one is searching for two things: simplicity and approximation.

Simplicity. One hopes that restriction of an operator to an invariant subspace will result in a simpler operator that provides insight into the workings of the original one. This is just what happens in the finite dimensional setting where the study of invariant subspaces leads to Schur's Theorem (Theorem 8.3 below), which asserts that every operator on $\mathbb{C}^N$ has—relative to an appropriately chosen orthonormal basis— an upper-triangular matrix. Schur's Theorem in turn leads to the Jordan Canonical

Form (see, e.g., [51, Chap. 3]), which tells us that every operator on $\mathbb{C}^N$ is similar to either an operator of the form $\lambda I + N$, where λ is a scalar and N is nilpotent (possibly the zero-operator), or to a direct sum of such operators.

For the infinite dimensional situation, suppose we have an operator T on a separable Hilbert space and that T has a nontrivial invariant subspace M. Upon choosing an orthonormal basis for M and completing it to one for the whole space we can write—just as in the finite dimensional case—a matrix (an infinite one this time) representing T with respect to this basis. This matrix will have a "block upper triangular" form $\left[\begin{smallmatrix} A & B \\ 0 & C \end{smallmatrix}\right]$, where the matrix A represents the restriction of T to M, B the restriction of PT to $M^\perp$ (P being the orthogonal projection of our Hilbert space onto M), and C the restriction to $M^\perp$ of $(I - P)T$. In fact the existence of a nontrivial invariant subspace is *equivalent* to T having such a matrix representation.

Approximation. For an operator T on a Banach space X, here's a natural way to construct an invariant subspace. Fix a non-zero vector $x_0 \in X$ and take the linear span of its iterate sequence under T, i.e., look at the linear subspace of X consisting of all vectors $p(T)x_0$, where p is a polynomial with complex coefficients. This linear subspace is taken into itself by T, hence so is its closure $\mathscr{V} = \mathscr{V}(T, x_0)$. Since $\mathscr{V}$ contains x_0, it is not the zero subspace; in fact $\mathscr{V}$ is the smallest T-invariant subspace containing x_0. If $\mathscr{V} \neq X$ then we've produced a nontrivial invariant subspace for T. On the other hand, if $\mathscr{V} = X$ then we have an approximation theorem: every vector in X is the limit of a sequence of polynomials in T applied to the *cyclic vector x_0*.

Example. Let T denote the linear transformation of "multiplication by x" on the Banach space $C([0, 1])$. More precisely,

$$(Tf)(x) = xf(x) \qquad (f \in C([0, 1]), 0 \leq x \leq 1).$$

It's easy to see that T is a bounded operator on $C([0, 1])$.

Claim: *The constant function 1 is a cyclic vector for T.*

Proof of Claim. For p a polynomial with complex coefficients, the vector $p(T)1$ is just p, now viewed as a function on $[0, 1]$. Thus $\mathscr{V}(T, 1)$ is the closure in $C([0, 1])$ of the polynomials. Now convergence in $C([0, 1])$ is uniform convergence on $[0, 1]$ so by the Weierstrass Approximation Theorem [101, Theorem 7.26, p. 159], $\mathscr{V}(T, 1) = C([0, 1])$. □

The operator of "multiplication by x" also makes sense for the Hilbert space $L^2([0, 1])$, and since the continuous functions are dense therein, the function 1 is a cyclic vector in that setting too. This is not to say that our operator T is devoid of nontrivial invariant subspaces; it has non-cyclic vectors, too. For example, in the setting of $C([0, 1])$ each function f that takes the value zero somewhere on $[0, 1]$ is a non-cyclic vector (exercise), so $\mathscr{V}(T, f)$ is a nontrivial invariant subspace for T.

Exercise 8.2. Characterize the cyclic vectors for the operator of "multiplication by x" when setting is

(a)   The Banach space $C([0,1])$.

(b)   The Hilbert space $L^2([0,1])$.

Exercise 8.3. Show that every operator on a *non-separable* Banach space has a nontrivial invariant subspace. Thus the invariant subspace problem really concerns only *separable* Banach spaces.

Exercise 8.4 (Reducing subspaces). A subspace is said to *reduce* an operator if it's invariant and has an invariant complement, i.e., if the whole space can be decomposed as the direct sum of the original invariant subspace and another one. Not every operator, even in finitely many dimensions, has a nontrivial reducing subspace; show that the operator induced on $\mathbb{C}^2$ by the matrix $\begin{bmatrix} 0 & 1 \\ 0 & 0 \end{bmatrix}$ does not have such a subspace. More generally the same is true for every $N \times N$ matrix whose N-th power is the zero-matrix, but whose $(N-1)$-st power is not.

Invariant subspaces and projections. Suppose X is a vector space, V a linear subspace, and P a projection taking X onto V, i.e., P is a linear transformation with $P(X) = V$ whose restriction to V is the identity operator.[1] The fact that P is the identity map when restricted to its image can be expressed by the equation $P^2 = P$. Clearly the linear transformation $Q = I - P$ is also a projection with $PQ = QP = 0$. Since $P + Q = I$ these equations tell us that the projections P and Q decompose X into the direct sum of $V = P(X)$ and $W = Q(X)$.

Proposition 8.2. *Suppose X is a vector space, V a linear subspace, P a projection taking X onto V, and $T : X \to X$ a linear transformation on X. Then the following three statements are equivalent:*

(a) $T(V) \subset V$.

(b) $PTP = TP$.

(c) $QTQ = QT$, where $Q = I - P$.

Proof. Statements (a) and (b) both assert that the restriction of P to $T(V)$ is the identity map on $T(V)$. As for the equivalence of (b) and (c): note that since $Q = I - P$ we have

$$QTQ = T - TP - PT + PTP = QT + (PTP - TP)$$

so $QTQ = QT$ if and only $PTP - TP = 0$.   □

8.2 Invariant Subspaces in $\mathbb{C}^N$

Invariant subspaces are important even for finite dimensional operators. For example, the following 1909 result of Issai Schur is a fundamental result in matrix theory.

---

[1] If X were a normed linear space with P continuous we could use the language introduced in Sect. 4.1 and call P a *retraction* of X onto V.

Theorem 8.3 (Schur's Triangularization Theorem). *Suppose V is a finite dimensional complex inner-product space and T is a linear transformation on V. Then V has an orthonormal basis relative to which the matrix of T is upper triangular.*

Schur's Theorem is really a statement about invariant subspaces. Suppose $\dim V = N$, and let $\mathscr{V} = (v_j : 1 \leq j \leq N)$ be the orthonormal basis it promises for the operator T (it's important to note here that "basis" means: "linearly independent spanning set, *written as an ordered list*"). Let $[T]$ denote the matrix of T with respect to this basis, i.e., for each index j, the j-th column of $[T]$ is the column vector of coefficients of Tv_j with respect to the basis $\mathscr{V}$. Thus the upper-triangularity of $[T]$ asserts that Tv_j belongs to the linear span V_j of the basis vectors $(v_1, v_2, \ldots, v_j)$, so for each j between 1 and N:

V_j is a nontrivial invariant subspace for T.

Schur's Theorem therefore promises, for each operator T on V, the existence of a descending chain of invariant subspaces

$$V = V_N \supset V_{N-1} \supset \cdots \supset V_1 \supset V_0 = \{0\}, \tag{8.1}$$

each of which has codimension one in the preceding one. It's an easy exercise to see that the existence of such a chain is equivalent to that of the basis promised by Schur's Theorem.

Proof of Schur's Theorem. This proceeds by induction on the dimension N. For $N = 1$ the theorem is trivial, so suppose $N > 1$ and the result is true for $N - 1$. The transformation T has an eigenvalue; let v_1 be a unit eigenvector for this eigenvalue, let V_1 be the (one dimensional) linear span of the singleton $\{v_1\}$, and let $W = V_1^\perp$, the orthogonal complement in V of V_1. The subspace W has dimension $N - 1$, but unfortunately it need not be invariant under T. To remedy this, let P denote the orthogonal projection of V onto W and consider the operator $R = PT$, for which W *is* invariant. Our induction hypothesis applies to the restriction $R|_W$ of R to W, and produces an orthonormal basis $(v_2, v_3, \ldots, v_N)$ for W relative to which the matrix of $R|_W$ is upper triangular.

Thus $(v_1, v_2, v_3, \ldots, v_N)$ is an orthonormal basis for V. We aim to show that the matrix of T with respect to this basis is upper triangular, i.e., that Tv_j lies in the linear span of the vectors $v_1, v_2, \ldots v_j$ for each index $1 \leq j \leq N$. We already know $Tv_1 \in V_1$, so suppose $j > 1$. We have

$$Tv_j = (I - P)Tv_j + PTv_j = (I - P)Tv_j + Rv_j$$

with $I - P$ the orthogonal projection of V onto V_1. Now R takes v_j into the subspace spanned by the vectors v_k for $2 \leq k \leq j$. Thus Tv_j belongs to the linear span of the vectors $(v_1, v_2, \ldots, v_j)$, as we wished to prove. □

Applications of Schur's Theorem. Before moving on let's see how Schur's Theorem makes short work of several fundamental results of linear algebra.

Hermitian Operators. Let V be a finite dimensional inner-product space, with inner product $\langle \cdot, \cdot \rangle$. Then to each operator T on V we can attach another one called the *adjoint T^** of T, defined by

$$\langle Tx, y \rangle = \langle x, T^*y \rangle \qquad (x, y \in V). \tag{8.2}$$

To say an operator T on V is *hermitian* means that $T = T^*$. If $(v_1, v_2, \ldots v_N)$ is an orthonormal basis for V, then an operator T on V is hermitian if and only if (8.2) holds with $T = T^*$ when x and y run through the elements of this basis, i.e., when

$$\langle Tv_i, v_j \rangle = \langle v_i, Tv_j \rangle = \langle Tv_j, v_i \rangle^*, \qquad (1 \le i, j \le N),$$

where the notation λ^*, when applied to a complex scalar λ, denotes "complex conjugate." Thus:

> *An operator T on V is hermitian if and only if, with respect to every (or even "some") orthonormal basis, its matrix and the conjugate-transpose of this matrix are the same.*

With these preliminaries in hand we obtain from Schur's Theorem—almost trivially—one of the most important theorems of linear algebra:

Corollary 8.4 (The Spectral Theorem for hermitian operators). *Suppose T is a hermitian operator on a finite dimensional inner-product space. Then the space has an orthonormal basis relative to which the matrix of T is diagonal.*

Proof. Schur's Theorem promises an orthonormal basis for the space, relative to which T has an upper-triangular matrix. With respect to this basis, the matrix of the adjoint T^* has all entries *above* the main diagonal equal to zero. But $T = T^*$, so the matrix of T has all entries off the main diagonal equal to zero. □

Why is this result is called a "spectral theorem?" For a finite dimensional operator, the set of eigenvalues is often called the "spectrum," and for each diagonal matrix this is precisely the set of diagonal entries. With this in mind, it's an easy exercise to check that the above Corollary can be restated:

> *If T is a hermitian operator on a finite dimensional inner-product space V then there is an orthonormal basis for V consisting of eigenvectors of T.*

Normal Operators. To say an operator on a finite dimensional inner-product space, or even a Hilbert space, is *normal* means that the operator commutes with its adjoint. Hermitian operators are normal, but not all normal operators are hermitian (Example: a diagonal matrix with at least one non-real entry.). It turns out that the spectral theorem for hermitian operators holds as well for normal operators. The proof follows the hermitian model, once we have the following surprisingly simple generalization of Schur's Theorem.

Theorem 8.5 (Schur's Theorem for Commuting Pairs of Operators). *If two operators commute on a finite dimensional inner-product space then the space has an orthonormal basis with respect to which each operator has upper-triangular matrix.*

Proof. This one is a minor modification of the induction proof of Theorem 8.3. Let V be the inner-product space in question, with $N = \dim V$, and let S and T be operators on V that commute. The result we want to prove is trivially true for $N = 1$, so suppose it holds for dimension $N - 1$, where $N > 1$. We want to prove it for dimension N. Once again we observe that T has an eigenvalue—call it μ, but now, instead of choosing just one unit T-eigenvector for μ, we look at the full eigenspace $E = \ker(T - \mu I)$, and note that since S commutes with T, Theorem 8.1 guarantees that this eigenspace is invariant for S. Thus the restriction of S to E has an eigenvalue λ, hence a corresponding unit eigenvector v_1, which by design is a λ-eigenvector for T. As before, let V_1 be the span of the single vector v_1, let $W = V_1^\perp$, and let P be the orthogonal projection of $\mathbb{C}^N$ onto W. Let $A = PS$ and $B = PT$. Both operators take W into itself, so if we can show that their restrictions to W commute, our induction hypothesis will supply an orthonormal basis for W relative to which the matrices of these restrictions are both upper triangular. Upon adjoining v_1 to this basis, then applying to both S and T the argument that finished off the proof of Theorem 8.3, we'll be done.

In fact, it's easy to see that A commutes with B on all of V. Since $W^\perp = V_1$ is invariant for both S and T, we know from the equivalence of (a) and (c) in Proposition 8.2 (with the roles of P and Q reversed) that $PTP = PT$ and $PSP = PS$. Thus

$$AB = PSPT = PST = PTS = PTPS = BA,$$

where the third inequality uses the commutativity of S and T.   □

Corollary 8.6 (The Spectral Theorem for normal operators on $\mathbb{C}^N$). *If T is a normal operator on $\mathbb{C}^N$ then there exists an orthonormal basis for $\mathbb{C}^N$ relative to which the matrix of T is diagonal.*

The above proof of Schur's Theorem for commuting pairs of operators can easily be extended to finite collections of commuting operators. The following exercise shows that this proof extends even further:

Exercise 8.5 (Triangularization for commuting families). Show that: *If $\mathscr{C}$ is a family of commuting operators on a finite dimensional inner-product space V, then there exists an orthonormal basis of V relative to which each operator in $\mathscr{C}$ has upper-triangular matrix.*

In particular, if the commuting family $\mathscr{C}$ consists of normal operators, there's an orthonormal basis for V relative to which each operator in the family has a diagonal matrix.

Outline of proof: The key is to prove that the family $\mathscr{C}$ has a common eigenvector; then the proof can proceed like that of Theorem 8.5. Note first that there are nontrivial subspaces of V that are $\mathscr{C}$-invariant (meaning: "invariant for every operator in $\mathscr{C}$"). Example: the eigenspace of any operator in $\mathscr{C}$. Let m be the minimum of the dimensions of all the eigenspaces of operators in $\mathscr{C}$, so $m \geq 1$. Choose a $\mathscr{C}$-invariant subspace of $\mathbb{C}^N$ having this minimum dimension m. Show that every operator in $\mathscr{C}$ is, when restricted to that subspace, a scalar multiple of the identity.

8.3 Compact Operators

The result we seek to understand, Lomonosov's Theorem, deals with two concepts: invariant subspaces and compact operators. Having spent some time getting a feeling for the former, let's now take a moment to review some of the fundamental facts about the latter.

A linear transformation on a normed linear space is said to be *compact* if it takes the closed unit ball into a relatively compact set. Since relatively compact sets are bounded it follows from Proposition C.8 (Appendix C, p. 196) that: *Every compact linear transformation is continuous,* and so is an "operator."

> *Exercise* 8.6 (Basic Facts About Compact Transformations). Here all linear transformations act on a normed linear space X.
>
> (a)   If $\dim X < \infty$ then every linear transformation on X is compact.
>
> (b)   For operators A and K on X: if K is compact then so are AK and KA (i.e., the compact operators on X form a closed ideal in the algebra of all operators).

The following exercise gives some feeling for the concept of compactness for a natural class of concrete operators on the Hilbert space ℓ^2.

> *Exercise* 8.7. For a bounded sequence $\Lambda := (\lambda_k)$ of complex numbers, define the linear "diagonal map" D_Λ on ℓ^2 by $D_\Lambda(x) = (\lambda_k \xi_k)$ for each vector $x = (\xi_k) \in \ell^2$. Show that D_Λ is continuous on ℓ^2, and compact if and only if $\lambda_k \to 0$.
>
> *Suggestion.* For compactness: first show that a subset S of ℓ^2 is relatively compact if and only if it is "equicontinuous at ∞" in the sense that
> $$\lim_{n \to \infty} \sup_{f \in S} \sum_{k \geq n} |f(k)|^2 = 0.$$

As noted in Exercise 8.6(a), every operator on a finite dimensional normed linear space is compact. By contrast we pointed out at the beginning of Sect. 7.2 that the unit ball of an infinite dimensional Hilbert space is *not* compact; according to Exercise 7.2 the same is true for $C([0,1])$. Thus the identity operator is not compact on either of these spaces. More is true:

Theorem 8.7. *If a normed linear space is infinite dimensional then its closed unit ball is not compact.*

This result, along with Proposition C.9 of Appendix C, shows that for normed linear spaces, compactness of the closed unit ball *characterizes* finite dimensionality. The key to its proof is the following lemma:

Lemma 8.8. *Suppose X is a normed linear space, Y a finite dimensional proper subspace, and $0 < r < 1$. Then there exists a unit vector $x \in X$ whose distance to Y is greater than r.*

Proof. Fix a vector $x_0 \in X$ that is not in Y, and let d denote the distance from x_0 to Y, i.e.,

$$d = \inf\{\|x_0 - y\| : y \in Y\}.$$

According to Corollary C.10 of Appendix C, the subspace Y is complete in the norm-induced metric on X, thus Y is closed in X. It follows that $d > 0$, hence there exists $y_0 \in Y$ with $\|x_0 - y_0\| < d/r$.

Claim: The unit vector $x = \frac{x_0 - y_0}{\|x_0 - y_0\|}$ is the one we seek.

Proof of Claim. If $y \in Y$ then

$$x - y = \frac{1}{\|x_0 - y_0\|}\left[x_0 - \underbrace{(y_0 + \|x_0 - y_0\|y)}_{\in Y}\right].$$

Thus the term on the right in square brackets has norm $\geq d$, so

$$\|x - y\| \geq d/\|x - x_0\| > d/(d/r) = r,$$

hence the distance from x to Y is $> r$, as desired.   □

Proof of Theorem 8.7. Let X be an infinite dimensional normed linear space. Fix a countable linearly independent set $\{x_n\}_1^\infty$ in X and let Y_n be the linear span of the vectors $\{x_1, \ldots, x_n\}$. There results the strictly increasing chain

$$Y_1 \subset Y_2 \subset Y_3 \subset \cdots$$

of subspaces of X, each of which is finite dimensional hence closed in its successor. By Lemma 8.8 there is, for each index $n > 1$, a unit vector $y_n \in Y_n$ at distance $\geq 1/2$ to y_{n-1}. Let $y_1 = x_1/\|x_1\|$. Suppose the indices i and j are different, say $i < j$. Then $y_i \in Y_{j-1}$, so $\|y_i - y_j\| \geq 1/2$. Thus (y_n) is a sequence of vectors in the closed unit ball of X that has no convergent subsequence.   □

Corollary 8.9. *The identity operator on a normed linear space is compact if and only if the space is finite dimensional.*

   This suggests that for operators, compactness should be intertwined with finite dimensionality. The following result gives one important way in which this is true; it's the beginning of what's known as "The Riesz Theory of Compact Operators."

Proposition 8.10. *Suppose K is a compact operator on a Banach space. If $\lambda \neq 0$ is an eigenvalue of K then the eigenspace $\ker(K - \lambda I)$ is finite dimensional.*

Proof. We may suppose without loss of generality that $\lambda = 1$ (exercise). Thus $M := \ker(K - I)$ is an invariant subspace for K and the restriction of K to M is a compact operator on M. Since this restriction equals the identity operator on M, the closed unit ball of M must be compact, hence M is finite dimensional by Theorem 8.7.   □

   On infinite dimensional Banach spaces, compact operators need not have eigenvalues. The exercise below provides an example: the Volterra operator, which was shown in Exercise 8.1 to have no eigenvalues.

Exercise 8.8 (Compactness Without Eigenvalues). Use the Arzela–Ascoli Theorem (Appendix B, Theorem B.8) to show that the Volterra operator is compact on $C([0,1])$.

8.4 Lomonosov's Theorem

We now turn to what is easily the most celebrated result on the existence of invariant subspaces. Here's a special case:

Theorem 8.11 (Lomonosov 1973). *Every non-zero compact operator on an infinite dimensional Banach space has a nontrivial hyperinvariant subspace.*

This result says that not only does every operator commuting with a non-zero compact have a nontrivial invariant subspace—already far more than was previously known—but also that there's even a nontrivial subspace invariant for *all* the operators commuting with the given compact. We'll devote the rest of this section to proving this remarkable result; the method of proof will provide an even more remarkable generalization.

The key to Theorem 8.11 is the following Lemma which, although Lomonosov did not state it explicitly, is in fact the crucial step in his argument.

Lemma 8.12. *Suppose X is an infinite dimensional Banach space and K is a non-zero compact operator on X. If K does* not *have a hyperinvariant subspace then there is an operator A on X that commutes with K and for which KA has a fixed point in $X \setminus \{0\}$.*

Proof that Lemma 8.12 implies Theorem 8.11. Suppose K is a non-zero compact operator on X that has no hyperinvariant subspace. Let A be as in the Lemma. Thus $M = \ker(AK - I)$ is not the zero subspace, and since AK is compact (by Exercise 8.6(b)) its eigenspace M is finite dimensional (Proposition 8.10), hence not equal to X. Now K commutes with A, hence it commutes with AK. Theorem 8.1 guarantees that M is invariant for every operator that commutes with AK, hence M is invariant for K. Since M is finite dimensional the restriction of K to M—hence K itself—has an eigenvalue; call it λ.

The corresponding eigenspace $E := \ker(K - \lambda I)$ is a non-zero subspace of X that is, by Theorem 8.1, invariant for every operator that commutes with K. Also, $E \neq X$; if $\lambda = 0$ this follows from the fact that $K \neq 0$, while if $\lambda \neq 0$ then it follows from the finite dimensionality of E. Thus E is a nontrivial hyperinvariant subspace for K, contradicting our assumption that K had no such subspace. Conclusion: K *does* have a nontrivial hyperinvariant subspace. □

Proof of Lemma 8.12. We're given a non-zero compact operator K on an infinite dimensional Banach space X and are assuming that K has *only trivial* hyperinvariant subspaces. Our goal is to produce an operator A that commutes with K such that AK has a non-zero fixed point (i.e., has 1 as an eigenvalue).

Step I. An Algebra of Operators. Let $\mathscr{A}$ denote the collection of operators on X that commute with K, the notation reflecting the fact that $\mathscr{A}$ is an *algebra* of operators, i.e., closed under addition, scalar multiplication and multiplication (= composition) of operators. In particular: for each $x \in X$ the set of vectors $\mathscr{A}x = \{Ax : A \in \mathscr{A}\}$ is a linear subspace of X (since $\mathscr{A}$ is closed under addition and scalar multiplication of operators) that's taken into itself by each operator in $\mathscr{A}$ (since $\mathscr{A}$ is closed under operator multiplication). Furthermore $\mathscr{A}$ contains the identity operator on X, so if $x \neq 0$ then $\mathscr{A}x \neq \{0\}$. Since we're assuming K has only trivial hyperinvariant subspaces, $\mathscr{A}x$ has to be dense for each $0 \neq x \in X$; otherwise its closure would be a nontrivial hyperinvariant subspace for K.

Step II. Some Sets. Since multiplication of K by a non-zero scalar changes neither its compactness, its commutation properties, nor its hyperinvariant subspace structure, we may without loss of generality assume that $\|K\| = 1$. Thus K is *contractive:* $\|Kx\| \leq \|x\|$ for every $x \in X$. Choose a vector $x_0 \in X$ for which $\|Kx_0\| > 1$. Because $\|K\| = 1$ this implies $\|x_0\| > 1$, so the closed ball

$$B = \{x \in X : \|x - x_0\| \leq 1\}$$

does not contain the origin. Let C denote the closure in X of $K(B)$. Since K is a compact operator and B is a bounded subset of X, the set C is compact. In addition, since B is convex and K linear, C is convex. Finally (and crucially), as the calculation below shows, C *does not contain the origin.* Indeed, for each $x \in X$:

$$\|Kx\| = \|K(x - x_0) + Kx_0\| \geq \|Kx_0\| - \|K(x - x_0)\| \geq \|Kx_0\| - \|x - x_0\|,$$

the last inequality arising from the contractivity of K. Thus for each $x \in B$ we have $\|Kx\| \geq \|Kx_0\| - 1 := \delta > 0$, hence every vector in $K(B)$, so also in its closure C, has norm at least δ.

> *Some wishful thinking.* If we could produce an operator $A \in \mathscr{A}$ for which $A(C) \subset B$, then KA, which also belongs to the algebra $\mathscr{A}$, would map the compact, convex set C continuously into itself, so by Schauder's theorem would have the desired fixed point. This is not quite what's going to happen, but it's still worth keeping in mind as we proceed.

Step III. A Map with a Fixed Point. Let B° denote the interior of the closed ball B. Suppose $0 \neq y \in X$. Since $\mathscr{A}y$ is dense in X there exists $A \in \mathscr{A}$ for which $y \in A^{-1}(B^\circ)$. Thus $\{A^{-1}(B^\circ) : A \in \mathscr{A}\}$ is an open cover of $X \backslash \{0\}$, hence an open cover of C. Since C is compact, it has a finite subcover $\mathscr{U} = \{U_j\}_1^N$, where $U_j := A_j^{-1}(B^\circ)$.

While we haven't produced a map $A \in \mathscr{A}$ with $A(C) \subset B$, we *have* produced a finite collection $\{A_1, A_2, \ldots, A_N\}$ of operators in $\mathscr{A}$, each of which takes *a piece of C into B*, as shown by the right-hand side of Fig. 8.1.

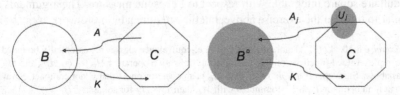

Fig. 8.1 What we want (*left*) vs. what we've got (*right*)

Lomonosov's great insight was to use a standard "nonlinear" argument to glue the operators A_j together into a continuous map that takes C into B. By Proposition B.6 of Appendix B there is a partition of unity subordinate to the open covering $\mathscr{U}$ of C, i.e., a set $\{p_j : 1 \le j \le N\}$ of continuous functions taking C into the unit interval $[0,1]$ that sum to 1 at each point of C, and have the property that for each index j the function p_j is $\equiv 0$ on $C\backslash U_j$. Define $\Phi : C \to X$ by

$$\Phi(y) = \sum_{j=1}^{N} p_j(y) A_j y \qquad (y \in C).$$

Being a finite sum of continuous maps, Φ is continuous. Moreover $\Phi(y)$ is, for each $y \in C$, a convex combination of vectors in the convex ball B, so it, too, belongs to B. Thus Φ is a continuous map taking C into B, hence $K \circ \Phi$ takes C continuously into itself. Since C is a compact, convex subset of a Banach space, the *Schauder Fixed-Point Theorem* (Theorem 7.1) guarantees that $K \circ \Phi$ has a fixed point $y_0 \in C$.

Step IV. Linearization. Let $A = \sum_{j=1}^{N} p_j(y_0) A_j$, a linear combination of operators in $\mathscr{A}$ and therefore also an operator in $\mathscr{A}$. Moreover

$$(KA)y_0 = K \left(\sum_{j=1}^{N} p_j(y_0) A_j y_0 \right) = K(\Phi(y_0)) = y_0.$$

Thus $A \in \mathscr{A}$ and $y_0 \in X\backslash\{0\}$ are the operator and vector we seek. This establishes Lemma 8.12 and with it, Lomonosov's Theorem 8.11.   □

Exercise 8.9. The hypothesis of Theorem 8.11 does not hold for every operator; there exist operators that commute with no non-zero compact operator. For $\varphi \in C([0,1])$ not identically zero, let M_φ denote the operator on $C([0,1])$ of "multiplication by φ," i.e.,

$$(M_\varphi f)(x) = \varphi(x) f(x) \qquad (0 \le x \le 1; f \in C([0,1])).$$

If $\varphi(x) \equiv x$ we'll write M_x instead of M_φ. Show that the operators M_φ are the only ones that commute with M_x, and that none of these is compact. *Hint:* If $T = M_\varphi$ then $\varphi = T1$.

We mentioned earlier that there are Banach space operators with no nontrivial invariant subspace, but that the problem is still open for Hilbert space (see the *Notes* at the end of this chapter for references and more details). Thus Exercise 8.9 would have more significance if it were set in a Hilbert space. The following modification does just that, replacing $C[0,1]$ with the Hilbert space $L^2 = L^2([0,1])$ consisting of (a.e.-equivalence classes of) measurable complex-valued functions on $[0,1]$ whose moduli are square integrable with respect to Lebesgue measure. The arguments are similar to those of the exercise above, but they require a bit more work.

Exercise 8.10. Let L^∞ denote the space of (a.e.-equivalence classes of) essentially bounded complex-valued functions on $[0,1]$. Define multiplication operators M_φ for $\varphi \in L^\infty$, and M_x, as above. Show that if $\varphi \in L^\infty\backslash\{0\}$ then M_φ is an operator on L^2 that is not compact. Show that if an operator T on L^2 commutes with M_x, then $T = M_\varphi$ for some $\varphi \in L^\infty$.

8.5 What Lomonosov *Really* Proved

According to Exercise 8.10 there exist operators on L^2 that commute with no non-zero compact operator. Consequently Lomonosov's Theorem 8.11, spectacular as it is, does not solve the Invariant Subspace Problem for Hilbert space. However the story does not end here. At the very end of his paper [71], Lomonosov notes that the reasoning he used to prove Theorem 8.11 yields more. In what follows, let's agree to call an operator "nonscalar" if it is not a scalar multiple of the identity operator.

Theorem 8.13 (Lomonosov). *If a nonscalar operator T on an infinite dimensional Banach space commutes with a non-zero compact operator, then T has a nontrivial hyperinvariant subspace.*

Our original Lomonosov Theorem implies that, on an infinite dimensional Banach space, every operator that commutes with a non-zero compact operator has a nontrivial invariant subspace. This one implies that a nontrivial invariant subspace exists for every operator that commutes with a nonscalar operator that commutes with a compact one.

Proof of Theorem 8.13. Let X be our infinite dimensional Banach space. The proof of Lemma 8.12 goes through word-for-word to establish this:

Lemma 8.12, Enhanced. *Suppose $\mathscr{A}$ is an algebra of operators on X, and K is a non-zero compact operator on X. Suppose there is no nontrivial closed subspace invariant for every member of $\mathscr{A}$. Then there exists an operator $A \in \mathscr{A}$ for which KA has a fixed point in $X \backslash \{0\}$.*

Suppose T is a nonscalar operator on X that commutes with our non-zero compact operator K. Let $\mathscr{A}$ denote the algebra of all operators that commute with T. We wish to show that there is a closed subspace, neither the zero subspace nor the whole space, that is invariant under every operator in $\mathscr{A}$. Suppose this is not the case. Then by the enhanced Lemma 8.12 we know that there exists $A \in \mathscr{A}$ such that KA has a fixed point in $X \backslash \{0\}$. The eigenspace $M := \ker(KA - I)$ is, just as before: $\neq \{0\}$, finite dimensional so $\neq X$, and invariant for every operator that commutes with KA. Since T commutes with both K and A, it commutes with KA, hence M is invariant for T. The restriction of T to the finite dimensional invariant subspace M therefore has an eigenvalue λ. The eigenspace $M_\lambda := \ker(T - \lambda I)$ is a closed subspace of X that is: not the zero subspace, not X (because T is nonscalar), and invariant for every operator that commutes with T. But we've assumed that $\mathscr{A}$ has no such subspace. Contradiction! Therefore $\mathscr{A}$ *does* have such a subspace. $\qquad\square$

Notes

Schur's Triangularization Theorem. This occurs in Schur's paper [109, p. 490], where it's applied to the study of integral equations. According to Horn and Johnson [51, p. 101], Schur's Theorem is "perhaps the most fundamentally useful fact of elementary matrix theory." Exercise 8.5 is from [51], see in particular Theorems 1.3.19, pp. 63–63 and 2.3.3, p. 103.

Compact operators. Lemma 8.8 is due to F. Riesz; it's Lemma 1 on p. 218 of his book [98] with Sz.-Nagy. Sections 76–80 of this book contain a nice exposition, set in the Hilbert space L^2, of the Riesz Theory of Compact Operators, a fundamental piece of which is—as we noted above—Proposition 8.10. The Riesz theory shows that compact operators behave "spectrally" very much like operators on finite dimensional spaces. For a modern exposition set in Banach spaces, see [103, Sects. 4.16–4.25, pp. 103–111]. J. H. Williamson showed in [124] that with the proper definition of "compact operator" the Riesz theory carries over to arbitrary (Hausdorff, but not necessarily locally convex) topological vector spaces.

Lomonosov's Theorem: prehistory. In the early 1930s von Neumann proved that every compact operator on Hilbert space has a nontrivial invariant subspace. He never published this result, and it was rediscovered about thirty years later by Aronszajn who, along with K. T. Smith, simplified the proof and in [4] generalized the result to Banach spaces.

The work of Aronszajn and Smith suggested the question of whether or not every operator whose *square* is compact has a nontrivial invariant subspace. This remained open until 1966 when Bernstein and Robinson in [11] showed, using non-standard analysis, that an operator has a nontrivial invariant subspace whenever some *polynomial* (not $\equiv 0$) in it is compact.

Various authors refined the Bernstein–Robinson proof, replacing their polynomial hypotheses with one of the form: "Some of limit of polynomials or rational functions in the operator is compact." Lomonosov's results superseded most of this earlier work. The version presented here of Lomonosov's work closely follows his original paper [71], as well as the exposition [92] of Pearcy and Shields.

Chains of commutation. For operators S and T on some Banach space, let's write $T \leftrightarrow S$ whenever S commutes with T, and let's write K for a generic non-zero compact operator. Theorem 8.11 implies that:

$$T \leftrightarrow K \implies T \text{ has a nontrivial invariant subspace.}$$

We've observed that, thanks to Exercise 8.10, the above consequence of Lomonosov's theorem doesn't solve the Invariant Subspace Problem for Hilbert space. However Theorem 8.13, the "real" Lomonosov Theorem, tells us that:

$$T \leftrightarrow S \text{ (nonscalar)} \leftrightarrow K \implies T \text{ has a nontrivial invariant subspace,}$$

so it makes sense to ask if *this* might solve the Invariant Subspace Problem for Hilbert space, i.e., "Does every operator on Hilbert space commute with a nonscalar operator that commutes with a non-zero compact?" This hope was destroyed in 1980 by Hadwin et al. [44].

One might still hope to solve the Invariant Subspace Problem by extending Lomonosov's method to get a result for longer "commutation chains." Unfortunately Troitsky in [119] showed that at least for the Banach space ℓ^1 there's no hope for such a result (see below for more details).

Counterexamples for Banach spaces. Per Enflo produced the first example of an operator on a Banach space possessing no nontrivial invariant subspace. Enflo's paper is [38, 1987], but his result was already circulating in preprint form over a decade earlier. A few years after Enflo released his preprint, Charles Read produced a much simpler counterexample, and then went on to find one set in the sequence space ℓ^1 [95, 1986]. Read later gave examples of Banach-space operators having no closed invariant *subset* [96, 1988].

In [119] Troitsky showed for Read's operator T on ℓ^1 that there exist non-scalar operators S_1 and S_2 on ℓ^1 such that $T \leftrightarrow S_1 \leftrightarrow S_2 \leftrightarrow K$, thus showing that Lomonosov's arguments cannot be extended to handle longer commutation chains.

In a totally different direction Argyros and Haydon [3] recently showed that there exist Banach spaces on which every bounded operator has the form "compact plus scalar multiple of the identity." Thus every bounded operator on such a space has a nontrivial invariant subspace (by the Aronszajn-Smith theorem), and even one that is *hyperinvariant* (by Lomonosov's theorem). Needless to say, such Banach spaces do not occur in the course of every-day mathematical life.

The current state of affairs. It's impossible to summarize quickly the many research efforts currently under way related to the Invariant Subspace Problem. The book [24, 2011] is an up-to-date exposition of the subject, while [94, 2003] is the standard reference for the state of the art circa 1973, and contains an outline, along with extensive references, of subsequent results up to 2003.

Part IV
Fixed Points for Families of Maps

In these final five chapters we'll turn our attention to fixed-point theorems involving, not just a single map, but a *family* of them for which we aim to produce a *common* fixed point. Necessarily we'll have to place severe restrictions on our classes of maps, but even so the results obtained will have surprising consequences that connect topology, algebra, and measure theory.

The fixed-point theorems we'll prove guarantee for every compact topological group the existence of Haar measure: a Borel probability measure that is invariant under the action of the group on itself. The model for this is arc-length measure on the unit circle, the group-invariance of which is the basis for Fourier analysis. The invariant measures we'll produce in the next few chapters play a similar role for the harmonic analysis of functions on compact groups, and we'll say something about how this goes in the abelian case.

An equally important thread involves the use of fixed-point theorems to produce *finitely additive* "measures" that are invariant under certain groups of transformations. This will lead us into the study of "paradoxical decompositions," the most famous example being the Banach–Tarski Paradox, which asserts that the unit ball of $\mathbb{R}^3$ can be split up into a finite number subsets that can be reassembled, using only rigid motions, into *two* unit balls. We'll spend some time understanding this paradox, and will show, via fixed-point theorems, that nothing similar is possible for either the unit circle or the unit disc.

Chapter 9
The Markov–Kakutani Theorem

FIXED POINTS FOR COMMUTING FAMILIES OF AFFINE MAPS

Overview. Consider the unit circle, the set $\mathbb{T}$ of complex numbers of modulus one. Complex multiplication makes $\mathbb{T}$ into a group, and the topology inherited from the complex plane makes it into a compact metric space. Here topology and algebra complement each other in that the group operations of multiplication $\mathbb{T} \times \mathbb{T} \to \mathbb{T}$ and inversion $\mathbb{T} \to \mathbb{T}$ are continuous. Tied up with the topology and algebra of $\mathbb{T}$ is arc-length measure defined on the Borel subsets of $\mathbb{T}$, the salient property of which is its *rotation invariance*: $\sigma(\gamma E) = \sigma(E)$ for each $\gamma \in \mathbb{T}$ and Borel subset E of $\mathbb{T}$.

In this chapter we'll study a remarkable fixed-point theorem due to Markov and Kakutani, based on which we'll show that not just the unit circle, but in fact *every* compact abelian group, has such a "Haar measure": a finite regular Borel probability measure invariant under the action of the group.[1] More generally, thanks again to the Markov–Kakutani theorem, we'll be able to produce both finitely and countably additive set functions that are invariant under quite general families of commuting transformations, a phenomenon that will point the way to our study in Chaps. 10–12 of the concepts of "amenability," "solvability," and "paradoxicality."

Prerequisites. Some general topology: bases, compactness, product topologies, continuity of mappings. Basic measure theory. Acquaintance with (or at least willingness to believe) the Tychonoff Product Theorem and the version of the Riesz Representation Theorem that produces measures from positive linear functionals.

9.1 Topological groups and Haar measure

Topological Groups. Suppose G is a group with its operation written multiplicatively. We'll think of group multiplication as a map $(x, y) \to xy$ that takes $G \times G$ into G, and inversion $x \to x^{-1}$ as a mapping of G into itself. If G has a topology (here,

---

[1] Haar measure is named for the Hungarian mathematician Alfred Haar (1885–1933). For further background see the *Notes* at the end of Chap. 13.

© Springer International Publishing Switzerland 2016
J.H. Shapiro, *A Fixed-Point Farrago*, Universitext,
DOI 10.1007/978-3-319-27978-7_9

always Hausdorff) that renders these two maps continuous, we'll call G, endowed with this topology, a *topological group*. Thus the circle group $\mathbb{T}$ described above is a compact topological group, and same is true of every product—both algebraic and topological—of $\mathbb{T}$ with itself. Euclidean space $\mathbb{R}^N$ with the usual topology and addition as its operation is a topological group that is not compact. Every group is a topological group in the discrete topology, the compact "discrete groups" being just the finite ones.

Exercise 9.1. Prove that:

(a)   The unit circle $\mathbb{T}$, as described above, is a topological group.

(b)   For each integer $N \geq 2$ the product space $\mathbb{T}^N$, consisting of N-tuples of elements of $\mathbb{T}$ is, with coordinatewise multiplication and the product topology (i.e., the topology it inherits from $\mathbb{C}^N$), a compact topological group.

(c) N-dimensional Euclidean space $\mathbb{R}^N$ is a topological group with its usual topology and the operation of vector addition.

(d) $GL_N(\mathbb{R})$, the collection of invertible $N \times N$ real matrices, endowed with the usual matrix operations and the topology it inherits as a subset of $\mathbb{R}^{N^2}$, is a (non-commutative) topological group.

The most commonly studied topological groups are the *locally compact* ones, i.e., those for which at every point the topology has a base of compact neighborhoods. All the examples in Exercise 9.1, indeed all the groups we'll study from now on, are locally compact. Except for occasional digressions, we'll focus our attention on the compact ones.

Exercise 9.2. Show that every infinite subgroup of the circle group $\mathbb{T}$ is dense. Use this result to show that the set of points $\{\sin n : n \in \mathbb{Z}\}$ is dense in the closed unit interval.

Borel sets and measures. In a topological space the collection of *Borel sets* is the sigma algebra generated by the open sets. Since sigma algebras are closed under the taking of complements and countable unions, each closed subset is a Borel set, as are countable unions and intersections of Borel sets.

Exercise 9.3 (Borel sets and continuity). Show that every continuous real-valued function on a topological space is measurable with respect to the Borel subsets of that space. Show that, at least for metric spaces, the sigma algebra of Borel sets is the smallest one with this property. Can you generalize this result beyond metric spaces?[2]

A *Borel measure* is simply a measure on the Borel sets of a topological space. To say a Borel measure is *regular* means that for every Borel set E:

$$\mu(E) = \inf\{\mu(U): U^{\text{open}} \supset E\} = \sup\{\mu(K): K^{\text{compact}} \subset E\} \qquad (9.1)$$

i.e., the measure of each Borel set can be approximated arbitrarily closely from the outside by open sets and from the inside by compact ones.

---

[2] For more on this see the Notes at the end of this chapter.

In this chapter we'll consider only regular Borel measures that are positive and have total mass one, i.e., *regular Borel probability measures* (henceforth: RBPMs).

Definition 9.1. A *Haar measure* for a compact topological group G is an RBPM that is *invariant* under the group action in the sense that $\mu(gB) = \mu(B)$ for every $g \in G$ and Borel subset B of G (here gB is the set of elements gb as b runs through B).

It turns out that every compact group has a (unique) Haar measure. In this chapter and the following two we'll use fixed-point theorems to prove this, concentrating for simplicity on the metrizable case. We'll discuss how these arguments can be enhanced to work in the general case, and in Chap. 12 will discuss an extension to locally compact groups.

Some examples of Haar measure. Arc-length measure (divided by 2π) for the unit circle $\mathbb{T}$, the product of arc-length measure (over 2π) with itself N times on $\mathbb{T}^N$, Lebesgue measure on $\mathbb{R}^N$.

> *Exercise 9.4.* Show that (commutative or not) every *finite group,* in its discrete topology, has a unique Haar measure.

> *Exercise 9.5.* Suppose G is a metrizable compact group with Haar measure μ. Show that if E is a Borel subset of G with $\mu(E) > 0$ then $E \cdot E^{-1}$ (the set of points xy^{-1} with x and y in E) contains an open ball.

> *Suggestion:* Show that the function $F \colon G \to [0, 1]$ defined by

$$F(x) = \int_G \chi_E(x^{-1}t)\chi_E(t)\,d\mu(t) \quad (x \in G)$$

> is continuous on G and not identically zero (the metrizability of G is not really needed; it's there to simplify the proof of continuity for the integral).

Left vs. right Haar measure. For non-commutative compact groups what we've been calling Haar measure should more accurately be called "left Haar measure," to distinguish it from "right Haar measure," i.e., a regular Borel probability measure μ for which $\mu(Bg) = \mu(B)$ for each Borel set B and group element g. We'll see in Chap. 12 (Theorem 12.15) that for compact groups the two concepts are the same and that Haar measure is *unique*, but that the situation for non-compact groups is more complicated; see Exercise 12.6.

9.2 Haar Measure as a Fixed Point

Measures and Functionals. To each finite regular Borel measure μ on a compact Hausdorff space Q there is an associated linear functional Λ_μ defined on $C(Q)$ (the space of continuous, real-valued functions on Q) by

$$\Lambda_\mu(f) = \int f\,d\mu \quad (f \in C(Q)).$$

If μ is a positive measure then the linear functional Λ_μ is *positive*: it takes non-negative values on functions having only non-negative values. Everything we do from now on will depend upon the following famous result, which asserts that such Λ_μ's are the *only* positive linear functionals on $C(Q)$.

The Riesz Representation Theorem for Compact Spaces.[3] *If Q is a compact topological space and Λ is a positive linear functional on $C(Q)$ then there is a unique positive regular finite Borel measure μ on Q such that $\Lambda = \Lambda_\mu$.*

Regularity is important here; If Q is a nasty enough compact space, a positive linear functional on $C(Q)$ may also be represented by a non-regular Borel probability measure (see, for example, [101, Chap. 2, Exercise 18, p. 59]). The good news: as shown by the exercise below, this can't happen for the most commonly occurring compact spaces.

Exercise 9.6. Show that for a compact metric space, every finite, positive Borel measure is regular.

Suggestion: Show that for such a measure μ, the collection of subsets that satisfy condition (9.1) above (i.e., the μ-regular sets) form a sigma algebra that contains all the closed sets.

Invariance via Functionals. For a compact topological group G (not necessarily commutative) and an RBPM μ on the Borel subsets of G, what property of Λ_μ corresponds to (left) G-invariance for μ?

Suppose μ is an RBPM for G. Then by the change-of-variable formula of measure theory:

$$\int f(\gamma x)\,d\mu(x) = \int f(x)\,d\mu\gamma^{-1}(x) \qquad (\gamma \in G, f \in C(G)), \tag{9.2}$$

where $\mu\gamma^{-1}$ is the measure that gives the value $\mu(\gamma^{-1}E)$ to the Borel subset E of G. Since G-invariance for μ just means that $\mu = \mu\gamma^{-1}$ for each $\gamma \in G$, Eq. (9.2) asserts that μ is G-invariant if and only if

$$\int f(\gamma x)\,d\mu(x) = \int f(x)\,d\mu(x) \tag{9.3}$$

for every $f \in C(G)$ and $\gamma \in G$. In order to rephrase this formula in terms of the linear functional Λ_μ, let's define for each $\gamma \in G$ the linear transformation $L_\gamma: C(G) \to C(G)$ of (left) *translation by γ*:

$$(L_\gamma f)(x) = f(\gamma x) \qquad (f \in C(G)). \tag{9.4}$$

In terms of the maps L_γ, the change-of-variable formula (9.2) becomes

$$\Lambda_\mu \circ L_\gamma = \Lambda_{\mu\gamma^{-1}} \qquad (\gamma \in G) \tag{9.5}$$

---

[3] See [101, Theorem 2.14, pp. 40–41], where the theorem is proved for *locally compact* spaces.

for each RBPM μ for G, while the invariance characterization (9.3) emerges as

$$\Lambda_\mu \circ L_\gamma = \Lambda_\mu \qquad (\gamma \in G). \tag{9.6}$$

With these observations we're one step away from being able to express an invariant measure as a fixed point. Here's the step.

Definition 9.2 (Dual space, adjoint). Let V be a real vector space and $T : V \rightarrow V$ a linear transformation.

(a) Denote by $V^\sharp$ the *algebraic dual* of V, i.e., the vector space of all linear functionals (linear transformations $V \rightarrow \mathbb{R}$) on V.

(b) Define the *adjoint* $T^\sharp$ of T by: $T^\sharp \Lambda = \Lambda \circ T$ for $\Lambda \in V^\sharp$.

One checks easily that $T^\sharp$ is a linear transformation $V^\sharp \rightarrow V^\sharp$. With these definitions the general transformation formula (9.5) becomes

$$L_\gamma^\sharp \Lambda_\mu = \Lambda_{\mu\gamma^{-1}} \qquad (\gamma \in G), \tag{9.7}$$

while the invariance condition (9.6) can be written

$$L_\gamma^\sharp \Lambda_\mu = \Lambda_\mu \qquad (\gamma \in G). \tag{9.8}$$

In summary:

Proposition 9.3. *An RBPM μ on a compact group G is (left) G-invariant if and only if its associated linear functional Λ_μ is a fixed point for each left-translation adjoint operator $L_\gamma^\sharp : C(G)^\sharp \rightarrow C(G)^\sharp$ $(\gamma \in G)$.*

9.3 The Markov–Kakutani Fixed-Point Theorem

Having translated the problem of finding Haar measure for a compact group into one of finding a fixed point for a family of linear maps, let's now turn our attention to a theorem that will guarantee the existence of such a fixed point. It turns out that some seemingly severe restrictions have to be made.

Commutativity. Our discussion of Haar measure began with the family of left-translation maps acting on the vector space $C(G)$ of continuous real-valued functions on the compact group G, then moved on to the family of adjoints of these maps acting on the algebraic dual space $C(G)^\sharp$. If G is commutative then it's easy to check that both families of maps—the translations and their adjoints—inherit (under composition) the commutativity of G. Now commutativity is a natural condition to impose upon a family of maps for which one hopes to find a common fixed point; it's an easy exercise to check that if a family of self-maps of some set commutes, then the set of fixed points of each map gets taken into itself by all the others. In particular, if one of the maps has a *unique* fixed point (e.g., if it's a strict contraction of a complete metric space) then that's a common fixed point for the whole family.

However, as the example below shows, a commutative family of maps, each of which has a fixed point, need not have a common fixed point—even if the maps are all continuous on a compact metric space.

Example 9.4. Let $S = \{1,2,3,4,5\}$ and $\Phi = \{\varphi, \psi\}$ where φ fixes 3, 4, and 5, and interchanges 1 and 2, while ψ fixes 1 and 2, and takes 3 to 4, 4 to 5, and 5 to 3. In the notation and language of permutations: φ is the 2-cycle [1 2] (also called a "transposition"), ψ is the 3-cycle [3 4 5], and being "disjoint" cycles, φ and ψ commute under composition. Thus S is compact in the discrete metric and Φ is a commuting family of continuous maps, each of which has a fixed point but for which there is no common fixed point.

Affine maps. Example 9.4 above shows that for a family of self-maps of a topological space: continuity plus commutativity plus compactness is still not enough to insure a common fixed point. What extra condition can we add to remedy this situation? Recall that in Sect. 9.2 above we found that the problem of existence Haar measure on a compact group is equivalent to that of finding a common fixed point for a family of *linear* maps. It turns out that if we add to the hypotheses of continuity, compactness, and commutativity, additional conditions of convexity and "affine-ness" then common fixed points do exist.

Definition 9.5 (Affine map). Suppose V is a real vector space, C a convex subset of V, and f is a map taking C into V. To say f is *affine* means that

$$f(tx + (1-t)y) = tf(x) + (1-t)f(y)$$

whenever $x, y \in C$ and $0 \le t \le 1$.

Restrictions of linear maps to convex sets are affine; these are the only affine maps we'll consider here.

Exercise 9.7. Suppose V is a real vector space. Show that:

(a)   If L is a linear map on the real vector space V and w is a vector in V, then the map $v \to Lv + w$ is affine on V.

(b)   The image of a convex subset of V under an affine map is again convex.

(c)   Affine mappings of convex subsets C of V respect convex combinations, i.e., for all n-tuples of vectors $(x_i : 1 \le i \le n)$ in C and non-negative scalars $(t_i : 1 \le i \le n)$ that sum to 1,

$$f\left(\sum_{i=1}^{n} t_i x_i\right) = \sum_{i=1}^{n} t_i f(x_i).$$

Vector Topology. The algebraic setting for our fixed-point theorem will be quite restrictive: commutative families of affine maps. By contrast the *topological* setting will be very general: (real) *topological vector spaces,* i.e., vector spaces V over the real field on which there is a topology (which we'll always require to be Hausdorff) that "respects" the vector operations. More precisely, the topology is required to render continuous[4]: addition, viewed as a map from the product space $V \times V$ into V, and scalar multiplication, viewed as a map $\mathbb{R} \times V \to V$. Such a topology is called a *vector topology*. For example, the norm-induced topology of a normed linear space is a vector topology; we'll soon discover others more suited to our purposes.

Exercise 9.8. Suppose U is a neighborhood of the zero vector in a topological vector space V. Show that $V = \bigcup_{n \in \mathbb{N}} nU$.

Hint: For each $x \in V$ the map $t \to tx$ ($t \in \mathbb{R}$) takes the real line continuously into V.

With this foundation we're now able to state the main result of this chapter.

Theorem 9.6 (The Markov–Kakutani Theorem). *Suppose V is a topological vector space inside of which K is a nonvoid compact, convex subset. Suppose $\mathscr{A}$ is a commutative family of continuous affine maps taking K into itself. Then there exists a point $p \in K$ such that $Ap = p$ for every $A \in \mathscr{A}$.*

Before proving the Markov–Kakutani Theorem, let's sketch how it might be used to produce Haar measure for a compact abelian group G. Continuing the discussion of Sect. 9.2: the vector space V of the theorem will be the algebraic dual $C(G)^{\sharp}$ of $C(G)$ and the convex set K will be the set of linear functionals Λ_{μ} on $C(G)$, where μ runs through the collection of RBPMs on G. The family $\mathscr{A}$ of affine self-maps of K will be the collection of *adjoints* $L_{\gamma}^{\sharp} \colon C(G)^{\sharp} \to C(G)^{\sharp}$ of the translation operators L_{γ} for $\gamma \in G$. As mentioned earlier: it's easily seen that $\mathscr{A}$ inherits the commutativity of G.

Equation (9.7) guarantees that each of the maps $L_{\gamma}^{\sharp}$ takes K into itself, so in order to apply the Markov–Kakutani theorem it remains to find a vector topology on $V = C(G)^{\sharp}$ rendering K compact and each $L_{\gamma}^{\sharp}$ continuous. Once this topology is found, the Markov–Kakutani Theorem will provide for $\mathscr{A}$ a fixed point in K and, as pointed out in Sect. 9.2, the Riesz Representation Theorem will provide the G-invariant RBPM corresponding to this fixed point. All this we'll do in Sect. 9.5. Right now, let's prove the fixed-point theorem.

9.4 Proof of the Markov–Kakutani Theorem

We'll break the proof into several pieces, the first being a straightforward consequence of the continuity of scalar multiplication. Throughout this section, V denotes a (real) topological vector space.

Lemma 9.7. *If K is a compact subset of V with $0 \in K$, then $\bigcap_{n \in \mathbb{N}} n^{-1}K = \{0\}$.*

Proof. Suppose U is a neighborhood of the zero vector in V. According to Exercise 9.8 above, the sets $\{nU : n \in \mathbb{N}\}$ cover V, so they cover K. Since K is compact

---

[4] "Continuity" of a map in this context means that the inverse image of any open set is open.

there is a finite subcover. Since the sets nU increase with n there exists $n \in \mathbb{N}$ such that $K \subset nU$, i.e., $n^{-1}K \subset U$. Thus $\bigcap_{n\in\mathbb{N}} n^{-1}K \subset U$ for each neighborhood U of zero in V. The desired result now follows from the fact that the topology of V is Hausdorff, so the intersection of all its zero-neighborhoods is $\{0\}$. $\qquad\square$

The next result is the heart of the Markov–Kakutani Theorem: the special case where the commuting family $\mathscr{A}$ consists of just a single map.

Proposition 9.8. *Suppose K is a compact, convex subset of V and A is an affine, continuous self-map of K. Then A has a fixed point in K. Moreover the set of all such fixed points is compact and convex.*

Proof. Let $\mathbb{N}^* = \mathbb{N} \cup \{0\}$, the set of non-negative integers. For $n \in \mathbb{N}^*$ let A^n denote the composition of A with itself n times (with A^0 denoting the identity map on K). Then each map A^n is an affine, continuous self-map of K, as is each *arithmetic mean* M_n defined by

$$M_n x = \frac{1}{n+1} \sum_{j=0}^{n} A^j x \qquad (x \in K, n \in \mathbb{N}^*).$$

Let $S = \bigcap_{n\in\mathbb{N}^*} M_n(K)$. Being an intersection of compact, convex sets, S is also compact and convex.

Claim. S is the fixed-point set of A.

Proof of Claim. Clearly every fixed point of A belongs to S. Conversely, fix $y \in S$. We wish to show that $Ay = y$. By the definition of S, for each $n \in \mathbb{N}^*$ there is a vector $x_n \in K$ such that $y = M_n x_n$. The map A, being affine, respects convex sums; in particular, $AM_n = M_n A$ for each n. Thus

$$Ay - y = AM_n x_n - M_n x_n = M_n A x_n - M_n x_n,$$

i.e.,

$$Ay - y = \frac{1}{n+1} \sum_{j=0}^{n} \left(A^{j+1} x_n - A^j x_n \right) = \frac{1}{n+1} \left(A^{n+1} x_n - x_n \right) \in \frac{1}{n+1}(K - K),$$

where $K - K$ is the set of all algebraic differences of pairs of elements of K. Since V is a topological vector space, the map $V \times V \to V$ defined by $(v, w) \to v + (-1)w$, is continuous, so $K - K$, the image under this map of the compact set K, is compact. In the above calculation n is an arbitrary non-negative integer, so $Ay - y$ belongs to $\bigcap_{n\in\mathbb{N}^*} \frac{1}{n+1}(K - K)$ which, by Lemma 9.7 above (and the fact that $0 \in K - K$), consists only of the zero vector. Thus $Ay = y$, as promised by the Claim.

So Far. We know that the compact, convex subset $S = \bigcap_{n\in\mathbb{N}^*} M_n(K)$ of K is the fixed-point set of A.

Remains to show. S is nonempty. To this end, let $\mathscr{M} = \{M_n : n \in \mathbb{N}^*\}$, so $\mathscr{M}(K) = \{M(K): M \in \mathscr{M}\}$ is a family of closed subsets of the compact set K, with $\bigcap \mathscr{M}(K) = S$. If we can show that each finite subfamily of $\mathscr{M}(K)$ has nonvoid intersection, then by the finite intersection property of compact sets, the same will be true of $\mathcal{M}(K)$ itself, thus finishing the proof.

Let $\mathcal{F}$ be a finite subfamily of $\mathcal{M}$ and let F be the composition of the maps $\mathcal{F}$, each map occurring exactly once in the composition. Since all the maps in $\mathcal{M}$ commute under composition (exercise), in the definition of F they can occur in any order. Thus for each $M \in \mathcal{F}$ we have $F = M \circ H$ where H is a self-map of K, hence $M(K) \supset M(H(K)) = F(K)$. Conclusion: $\bigcap_{M \in \mathcal{F}} M(K) \supset F(K) \neq \emptyset$. $\square$

Finally, we complete the proof of (the full-strength version of) the Markov–Kakutani Theorem.

Proof of Theorem 9.6. We're given a compact, convex subset K of the topological vector space V, and a family $\mathcal{A}$ of affine, continuous self-maps of K that commute under composition. Our goal is to show that there is a common fixed point for all the maps in $\mathcal{A}$.

For $A \in \mathcal{A}$ let $S_A = \{x \in K : Ax = x\}$, the fixed-point set of A. From Proposition 9.8 we know that S_A is a convex, compact subset of K that is not empty. We desire to show that $\bigcap_{A \in \mathcal{A}} S_A$, the common fixed-point set for $\mathcal{A}$, is nonempty. For this it's enough—again by the finite intersection property of compact sets—to show that

$$\bigcap_{A \in \mathcal{F}} S_A \neq \emptyset \qquad (*)$$

for each finite subfamily $\mathcal{F}$ of $\mathcal{A}$.

We proceed by induction on the number n of elements of $\mathcal{F}$, the case $n = 1$ being just Proposition 9.8. Suppose $(*)$ is true for some $n \geq 1$, and that $\mathcal{F}$ is a subfamily of $\mathcal{A}$ consisting of $n+1$ maps. Pick a map A out of $\mathcal{F}$ and let S denote the common fixed-point set of the n maps that remain. Then S, being the intersection of n compact, convex subsets of K, is again compact and convex in K; by our induction hypothesis $S \neq \emptyset$. By commutativity, $A(S_T) \subset S_T$ for each $T \in \mathcal{F} \setminus \{A\}$, hence A maps S, the intersection of these sets, into itself. By Proposition 9.8, A has a fixed point in S, which is therefore a common fixed point for $\mathcal{F}$. Conclusion: $(*)$ holds for each subfamily $\mathcal{F}$ consisting of $n+1$ maps, so by induction it holds for every finite subfamily of $\mathcal{A}$. $\square$

9.5 Markov–Kakutani Operating Manual

In order to enlist the Markov–Kakutani theorem in the production of invariant measures we need to find an appropriate vector topology for $C(G)^\sharp$, the algebraic dual space of $C(G)$. For this it's best to think of $C(G)$ as just a set, say S, and to view $C(G)^\sharp$ as a subspace of $\mathbb{R}^S$, the vector space of *all* real-valued functions on S. It's on this larger space that we'll define our vector topology.

Let $\omega(S)$ be the topology on $\mathbb{R}^S$ for which each $f \in \mathbb{R}^S$ has a base of neighborhoods defined as follows. For $\varepsilon > 0$ and F a finite subset of S, let

$$N(f,F,\varepsilon) = \{g \in \mathbb{R}^S : |g(s) - f(s)| < \varepsilon \quad \forall s \in F\}. \tag{9.9}$$

The following exercise shows that the collection of sets (9.9) really is a base for a topology on $\mathbb{R}^S$.

> *Exercise 9.9.* Show that if two sets $N(f_j, F_j, \varepsilon_j)$ $(j = 1, 2)$ have nonempty intersection, then that intersection contains a third set $N(f_3, F_3, \varepsilon_3)$.

$\omega(S)$ is the *product topology* (general definition given in the paragraph after Exercise 9.11 below) one obtains by viewing $\mathbb{R}^S$ as the topological product $\prod_{s \in S} \mathbb{R}_s$, where $\mathbb{R}_s = \mathbb{R}$ for each $s \in S$. It is often called the "topology of pointwise convergence on S." The next exercise explains why.

> *Exercise 9.10.* Show that a sequence of real-valued functions on S converges in the topology $\omega(S)$ if and only if it converges pointwise on S.

Proposition 9.9. *$\omega(S)$ is a vector topology on $\mathbb{R}^S$.*

Proof. The first order of business is to show that the topology $\omega(S)$ is *Hausdorff*. Given f_1 and f_2, distinct functions in $\mathbb{R}^S$, we want to find neighborhoods N_j of f_j $(j = 1, 2)$ with $N_1 \cap N_2 = \emptyset$. Since $f_1 \neq f_2$ there exists $s \in S$ for which $|f_1(s) - f_2(s)| = \varepsilon > 0$. Then $f_j \in N_j = N(f_j, \{s\}, \varepsilon/2)$ for $(j = 1, 2)$, and $N_1 \cap N_2 = \emptyset$.

It remains to establish *continuity* for the mappings $\sigma : \mathbb{R}^S \times \mathbb{R}^S \to \mathbb{R}^S$ and $\rho : \mathbb{R} \times \mathbb{R}^S \to \mathbb{R}^S$ of vector addition and scalar multiplication, defined respectively by

$$\sigma(f,g) = f + g \quad \text{and} \quad \rho(t,f) = tf \quad (f, g \in \mathbb{R}^S, t \in \mathbb{R}).$$

Continuity of σ. Suppose W is an open subset of $\mathbb{R}^S$. We need to show that $\sigma^{-1}(W) = \{(f_1, f_2) \in \mathbb{R}^S \times \mathbb{R}^S : f_1 + f_2 \in W\}$ is open in $\mathbb{R}^S \times \mathbb{R}^S$. Fix $(f_1, f_2) \in \sigma^{-1}(W)$ and choose $\varepsilon > 0$ and a finite subset F of $\mathbb{R}^S$ so that $N(f_1 + f_2, F, \varepsilon) \subset W$. Then $U := N(f_1, F, \varepsilon/2) \times N(f_2, F, \varepsilon/2)$ is an open subset of $\mathbb{R}^S \times \mathbb{R}^S$ that contains (f_1, f_2). One checks easily that $\sigma(U) \subset N(f_1 + f_2, F, \varepsilon) \subset W$, hence $U \subset \sigma^{-1}(W)$. Thus $\sigma^{-1}(W)$ is open in V, as desired.

Continuity of ρ. Fix $f_0 \in \mathbb{R}^S$ and $t_0 \in \mathbb{R}$. Suppose we're given $\varepsilon > 0$ and a finite subset F of $\mathbb{R}^S$. Our goal is to find an open interval N_1 about t_0 and an $\omega(S)$-neighborhood N_2 of f_0 such that $\rho(N_1 \times N_2) \subset N(t_0 f_0, F, \varepsilon)$. In plain language, we're looking for positive numbers δ_1 and δ_2, and a finite subset of $\mathbb{R}^S$—which can only be F itself—such that:

$$|t - t_0| < \delta_1 \quad \text{and} \quad |f - f_0| < \delta_2 \text{ on } F \implies |tf - t_0 f_0| < \varepsilon \text{ on } F.$$

An "epsilon-halves" argument (exercise) shows that we can get the desired result by setting $M = \max_{s \in F} |f_0(s)|$, then taking $\delta_1 = \frac{\varepsilon}{2M}$ and $\delta_2 = \frac{\varepsilon}{2(|t_0| + \delta_1)}$. $\qquad\square$

From Now on: We'll always assume $\mathbb{R}^S$ to be endowed with the topology $\omega(S)$.

Points as Functions. We can view each point s of a set S as a function $\hat{s} \colon \mathbb{R}^S \to \mathbb{R}$, where $\hat{s}(f) = f(s)$ for $f \in \mathbb{R}^S$. In the language of product spaces we can think of $\hat{s}$ as the "projection" of $\mathbb{R}^S$ onto its "s-th coordinate."

Proposition 9.10. *For each $s \in S$ the function $\hat{s} \colon \mathbb{R}^S \to \mathbb{R}$ is continuous on $\mathbb{R}^S$; the topology $\omega(S)$ is the weakest one for which this is true.*

Proof. For $t_0 \in \mathbb{R}$ and $\varepsilon > 0$ let I be the open interval of radius ε centered at t_0. Then $\hat{s}^{-1}(I) = N(f, \{s\}, \varepsilon)$ for all $f \in \mathbb{R}^S$ for which $f(s) = t_0$. Thus the inverse image under $\hat{s}$ of each real open interval is an open subset of $\mathbb{R}^S$, establishing the continuity of $\hat{s}$. Furthermore this argument shows that in every topology τ on $\mathbb{R}^S$ for which each of the functions $\hat{s}$ is continuous, $N(f, \{s\}, \varepsilon)$ has to be an open set, and since the basic open sets for $\omega(S)$ are finite intersections of these, every $\omega(S)$-open set must be τ-open, i.e., the topology τ must be at least as strong as $\omega(S)$. $\qquad\square$

Compactness in $\mathbb{R}^S$. The Markov–Kakutani Theorem requires compact sets. For finite dimensional normed linear spaces there are lots of these; the Heine–Borel Theorem asserts that every bounded subset of $\mathbb{R}^N$ has compact closure. However we saw in Theorem 8.7 that nothing of the sort can happen once the dimension of our normed space becomes infinite. Fortunately, our vector topology $\omega(S)$ on the space $\mathbb{R}^S$ turns out to be weak enough to allow the re-emergence of Heine–Borel-like phenomena. The key to this is the *Tychonoff Product Theorem,* which states that arbitrary topological products of compact spaces are compact.[5] In its full generality Tychonoff's theorem follows from the Axiom of Choice (Appendix E.3 below); for more on this see the *Notes* at the end of Appendix E.3.

Definition 9.11. To say a subset $\mathscr{B}$ of $\mathbb{R}^S$ is *pointwise bounded* means that for every $s \in S$, $\sup_{b \in \mathscr{B}} |b(s)| < \infty$ (i.e., the projection $\hat{s}$ is bounded on $\mathscr{B}$ for each $s \in S$).

Theorem 9.12 (A Heine–Borel Theorem for $\mathbb{R}^S$). *Let S be a set and $\mathscr{B}$ a subset of $\mathbb{R}^S$. Then $\mathscr{B}$ has compact closure in $\mathbb{R}^S$ if and only if it is pointwise bounded.*

Proof. (a) Suppose $\mathscr{B}$ is pointwise bounded. For $s \in S$ let $m_s = \sup_{b \in \mathscr{B}} |b(s)|$, and let I_s denote the compact real interval $[-m_s, m_s]$. Thus $\mathscr{B}$ is a subset of the product space $\mathscr{P} := \prod_{s \in S} I_s$, which is compact by the Tychonoff Product Theorem. Now $\mathscr{P}$ is a subset of $\mathbb{R}^S$; it's the collection of functions on $f \colon S \to \mathbb{R}$ for which $f(s) \in I_s$ for each $s \in S$. By its definition, the product topology on $\mathscr{P}$ is just the restriction to that set of the topology $\omega(S)$. Since $\mathscr{P}$ is compact in this topology, and $\mathscr{B} \subset \mathscr{P}$, the closure of $\mathscr{B}$ lies in $\mathscr{P}$ and so is compact.

(b) Suppose, conversely, that $\mathscr{B}$ has compact closure in $\mathbb{R}^S$. Each "projection" $\hat{s}$, being continuous on $\mathbb{R}^S$ (Proposition 9.10), is bounded an every compact subset, and in particular on $\mathscr{B}$, i.e., $\mathscr{B}$ is pointwise bounded. $\qquad\square$

Suppose now that V is a real vector space. Note that $V^\sharp$, the algebraic dual of V, is a vector subspace of $\mathbb{R}^V$.

---

[5] See Exercise 9.11 below for an accessible special case of this.

Definition 9.13. (The Weak-Star Topology.) The restriction to $V^\sharp$ of the product topology $\omega(V)$ on $\mathbb{R}^V$ is a vector topology that we'll call the *weak-star topology induced on $V^\sharp$ by V*.

The next result is crucial to the application of our infinite dimensional version of the Heine–Borel Theorem.

Proposition 9.14. $V^\sharp$ *is closed in $\mathbb{R}^V$.*

Proof. We need to show that every limit point of $V^\sharp$ in $\mathbb{R}^V$ belongs to $V^\sharp$. So suppose $\Lambda_0 \in \mathbb{R}^V$ is such a limit point; it's a real-valued function on V that we wish to prove is linear.

Λ_0 *is additive.* Fix x and y in V; we wish to show that $\Lambda_0(x+y) = \Lambda_0(x) + \Lambda_0(y)$. To this end let $\varepsilon > 0$ be given and consider the basic neighborhood $U := N(\Lambda_0, \{x, y, x+y\}, \varepsilon/3)$ of Λ_0. Since Λ_0 is a limit point of $V^\sharp$ this neighborhood contains an element Λ of $V^\sharp$. By the definition of U (Eq. (9.9), p. 110) we have $|\Lambda_0(w) - \Lambda(w)| < \varepsilon/3$ for $w \in \{x, y, x+y\}$. Thus

$$|\Lambda_0(x+y) - \Lambda_0(x) - \Lambda_0(y)|$$

$$= |\Lambda_0(x+y) - \Lambda(x+y) + [\Lambda(x) - \Lambda_0(x)] + [\Lambda(y) - \Lambda_0(y)]|$$

$$\leq \underbrace{|\Lambda_0(x+y) - \Lambda(x+y)|}_{<\varepsilon/3} + \underbrace{|\Lambda(x) - \Lambda_0(x)|}_{<\varepsilon/3} + \underbrace{|\Lambda(y) - \Lambda_0(y)|}_{<\varepsilon/3}$$

$$< \varepsilon.$$

Since ε is an arbitrary positive number, $\Lambda_0(x+y) - \Lambda_0(x) - \Lambda_0(y) = 0$, as desired.

Λ_0 *Is homogeneous.* Fix $t \in \mathbb{R}$ and $x \in V$; we wish to prove that $\Lambda_0(tx) = t\Lambda_0(x)$. Let $\varepsilon > 0$ be given; set $\delta = \varepsilon/2$ if $|t| \leq 1$, and $= \varepsilon/(2|t|)$ otherwise. As before, $N(\Lambda_0, \{x, tx\}, \delta)$ contains some $\Lambda \in V^\sharp$, so

$$|\Lambda_0(tx) - t\Lambda_0(x)| = |\Lambda_0(tx) \underbrace{-\Lambda(tx) + t\Lambda(x)}_{=0} - t\Lambda_0(x)|$$

$$\leq \underbrace{|\Lambda_0(tx) - \Lambda(tx)|}_{<\varepsilon/2} + |t| \underbrace{|\Lambda(x) - \Lambda_0(x)|}_{<\varepsilon/2}$$

$$< \varepsilon.$$

Thus (arbitrariness of ε once again) $\Lambda_0(tx) - t\Lambda_0(x) = 0$, as desired. $\qquad\square$

Corollary 9.15 (A Heine–Borel Theorem for Algebraic Duals). *For each real vector space V, a subset of $V^\sharp$ is weak-star compact if and only if it is weak-star closed and pointwise bounded.*

Proof. Suppose $\mathscr{E}$ is a weak-star compact subset of $V^\sharp$, hence closed therein. For each $w \in V$ the coordinate projection $\hat{w}\colon \Lambda \to \Lambda(w)$ is a continuous function $V^\sharp \to \mathbb{R}$, and so is bounded on $\mathscr{E}$. Thus $\mathscr{E}$ is pointwise bounded on V.

Conversely, suppose $\mathscr{E}$ is a weak-star closed subset of $V^\sharp$ that is pointwise bounded on V. By Theorem 9.12, $\mathscr{E}$ has compact closure in $\mathbb{R}^V$. By Proposition 9.14 the closure of $\mathscr{E}$ in $\mathbb{R}^V$ is the same as its closure in $V^\sharp$, which equals $\mathscr{E}$. Thus $\mathscr{E}$ is weak-star compact in $V^\sharp$. $\qquad\qquad\square$

Exercise 9.11 (A "proto-Tychonoff" theorem). Suppose S is a *countable* set, say (without loss of generality) $S = \mathbb{N}$. Define a function $\mathbb{R}^S \times \mathbb{R}^S \to [0,1]$ by:

$$d(f,g) = \sum_{n=1}^{\infty} \frac{1}{2^n} \frac{|f(n) - g(n)|}{1 + |f(n) - g(n)|} \ .$$

(a)   Prove that d is a metric on $\mathbb{R}^S$ and that $\omega(S)$ is the topology it induces thereon.

(b)   Use sequential arguments to prove that $\omega(S)$ is a vector topology on $\mathbb{R}^S$.

(c)   Use a diagonal argument to establish Theorem 9.12 for this special case.

Exercise 9.12. Suppose S is an *uncountable* set, and V is the set of functions in $\mathbb{R}^S$ whose zero-set is at most countable. Show that V is a vector subspace of $\mathbb{R}^S$ that is sequentially closed (i.e., every sequence in V that is $\omega(S)$-convergent has its limit in V), but not closed. In particular, the topology $\omega(S)$ is not metrizable.

Recall our original motivation for the topology $\omega(S)$. Given a compact abelian group G, we wished to apply the Markov–Kakutani Theorem to produce Haar measure for G. For this we needed to apply the theorem, not to the vector space $C(G)$, but rather to its *algebraic dual space* $C(G)^\sharp$. Thus we took the set S of the previous discussion to be $C(G)$ itself, with the idea of restricting the topology $\omega = \omega(C(G))$ to the dual space $C(G)^\sharp$. To complete our program we need to establish both the compactness of the convex set K of functionals Λ_μ where μ runs through the RBPMs on G, and the ω-continuity of the commutative family $\{L_\gamma^\sharp : \gamma \in G\}$ of adjoint self-maps of K.

Clarity through abstraction: Our arguments will be best understood in a more general setting. For this we'll replace our compact abelian group G by a nonempty set S, assumed to carry no topology at all. We'll replace the left-translation mappings furnished by the group operation with a commutative family Φ of self-maps of S. Finally, the role of the space $C(G)$ will be usurped by $B(S)$, the vector space of all functions $f\colon S \to \mathbb{R}$ that are *bounded*, i.e., for which

$$\|f\| := \sup\{|f(s)| : s \in S\} < \infty. \tag{9.10}$$

It's easy to check that $\|\cdot\|$ is a norm that makes $B(S)$ into a Banach space, but— perhaps surprisingly—we'll never need this fact.

The vector space V to which we'll apply the Markov–Kakutani Theorem will be $B(S)^\sharp$, endowed with the weak-star topology ω it inherits as a subspace of $\mathbb{R}^{B(S)}$. The compact, convex subset K of the Markov–Kakutani Theorem will be the set of "means" in $B(S)^\sharp$, defined as follows:

Definition 9.16. A *mean* is an element of $B(S)^\sharp$ that's positive (takes non-negative values on non-negative functions) and takes the value 1 at the constant function 1.

Notation. We'll use $\mathscr{M}(S)$ to denote the collection of means in $B(S)^\sharp$.

Exercise 9.13 (Evaluation functionals are means). For a set S:

(a)   Show that $\mathscr{M}(S)$ is a convex subset of $B(S)^\sharp$.

(b)   Show that for each $s \in S$ the *evaluation functional* $f \to f(s)$ is a mean (so $\mathscr{M}(S)$ is nonempty).

Exercise 9.14 (Means and mean values). Each number claiming to be a "mean value" for a bounded function should at least lie between the function's infimum and supremum. Show that $\Lambda(f)$ has this property for each $\Lambda \in \mathscr{M}(S)$ and $f \in B(S)$.

Exercise 9.15. Show that the convex hull of the evaluation functionals of Exercise 9.13(b)—a subset of $\mathscr{M}(S)$ by that exercise—*exhausts* $\mathscr{M}(S)$ if and only if S is a finite set.

Finally, the family of commuting affine maps for which we wish to find a common fixed point will be the adjoints of *composition operators* induced on $B(S)$ by the maps in Φ. More precisely, for each $\varphi \in \Phi$ define $C_\varphi : B(S) \to B(S)$ by $C_\varphi f = f \circ \varphi$. Clearly C_φ is a linear transformation $B(S) \to B(S)$ that preserves positivity and fixes the constant functions. We'll denote by $\mathscr{C}_\Phi$ the collection of all these maps, and $\mathscr{C}_\Phi^\sharp$ the collection of their *adjoints*. Once checks easily that $\mathscr{C}_\Phi^\sharp$ is commutative, and that each member of $\mathscr{C}_\Phi^\sharp$ takes $\mathscr{M}(S)$ into itself.

So Far: We've assembled the cast of characters demanded by the Markov–Kakutani Theorem, namely the vector space $V = B(S)^\sharp$, the commutative family of affine maps $\mathscr{A} = \mathscr{C}_\Phi^\sharp = \{C_\varphi^\sharp : \varphi \in \Phi\}$, the convex set $K = \mathscr{M}(S)$ on which these maps act, and the vector topology ω on $B(S)^\sharp$ that's going to glue these actors together. What's left is to show that $\mathscr{M}(S)$ is ω-compact and each of the maps $C_\varphi^\sharp$ is ω-continuous.

Corollary 9.17. $\mathscr{M}(S)$ *is ω-compact.*

Proof. By Corollary 9.15 it's enough to show that $\mathscr{M}(S)$ is pointwise bounded on $B(S)$ and ω-closed in $B(S)^\sharp$. By Exercise 9.14 we know for each $f \in B(S)$ and $\Lambda \in \mathscr{M}(S)$ that the value $\Lambda(f)$ lies between $\inf_{s \in S} f(s)$ and $\sup_{s \in S} f(s)$. Thus $\mathscr{M}(S)$ is a pointwise bounded subset of $B(S)$.

To show that $\mathscr{M}(S)$ is ω-closed in $B(S)^\sharp$, suppose $\Lambda_0 \in B(S)^\sharp$ is a limit point of $\mathscr{M}(S)$. We wish to show that $\Lambda_0 \in \mathscr{M}(S)$, i.e., that Λ_0, which we already know is a linear functional on $B(S)$, is *positive* and "normalized" so that $\Lambda_0(1) = 1$.

To establish positivity, fix $f \in B(S)$ with $f \geq 0$ on S. Let $\varepsilon > 0$ be given. Then the basic ω-neighborhood $N(\Lambda_0, \{f\}, \varepsilon)$ of Λ_0 contains a point $\Lambda \in \mathscr{M}(S)$. Thus

$$\Lambda(f) - \Lambda_0(f) \leq |\Lambda(f) - \Lambda_0(f)| < \varepsilon, \tag{9.11}$$

hence $\Lambda(f) \leq \Lambda_0(f) + \varepsilon$ for every $\varepsilon > 0$ so $\Lambda(f) \leq \Lambda_0(f)$. But $0 \leq \Lambda(f)$ since Λ, being a member of $\mathscr{M}(S)$, is a positive linear functional on $B(S)$. Conclusion: $0 \leq \Lambda_0(f)$, thus establishing the positivity of the limit functional Λ_0.

As for "normalization," choose $f \equiv 1$ on S. By the second inequality of (9.11): for each $\varepsilon > 0$, $|1 - \Lambda_0(f)| < \varepsilon$ hence $\Lambda_0(f) = 1$.   □

It remains to establish ω-continuity for the adjoint maps $C_\varphi^\sharp$. This follows from something more general:

Proposition 9.18 (Adjoints Are continuous). *If V is a real vector space and T a linear transformation of V into itself, then the adjoint map $T^\sharp$ is weak-star continuous on $V^\sharp$.*

Proof. In a topological vector space V, the map $x \to x + h$ of translation by a fixed vector h is a linear homeomorphism, so a linear transformation on V is continuous if and only if it is continuous at the origin. Thus it's enough to show that $T^\sharp$ is continuous at the origin of $V^\sharp$, and for this it's enough to show that the $T^\sharp$-inverse image of each basic zero-neighborhood in $V^\sharp$ contains a basic zero-neighborhood. For this, suppose $\varepsilon > 0$ and F is a finite subset of V. Upon chasing definitions one sees that $(T^\sharp)^{-1}(N(0, F, \varepsilon)) = N(0, T(F), \varepsilon)$.   □

Thus the Markov–Kakutani Theorem applies to the triple $(B(S)^\sharp, \mathscr{M}(S), \mathscr{C}_\Phi^\sharp)$; it yields the following:

Theorem 9.19 (Invariant means). *Suppose Φ is a commutative family of self-maps of a set S. Then there is a mean Λ in $B(S)^\sharp$ such that $C_\varphi^\sharp \Lambda = \Lambda$ (i.e., $\Lambda \circ \varphi = \Lambda$).*

9.6 Invariant Measures for Commuting Families of Maps

Theorem 9.19 yields, as a special case, the theorem that started our quest.

Corollary 9.20 (Haar Measure for compact abelian groups). *Every compact abelian group G supports, on its Borel subsets, a G-invariant RBPM.*

Proof. Apply Theorem 9.19 with $S = G$ and Φ the collection of translation-maps $x \to \gamma \cdot x$ for γ and x in G. The resulting composition operators on $B(G)$ are the translation operators L_γ on $B(G)$. For this situation Theorem 9.19 provides a mean Λ on $B(G)$ that's fixed by each of the transformations $L_\gamma^\sharp$ for $\gamma \in G$, so by the Riesz Representation Theorem and Proposition 9.3, the restriction of this functional to $C(G)$ has the form Λ_μ, where μ is an RBPM on the Borel subsets of G.   □

The argument above gives a more general result:

Corollary 9.21. *Suppose Q is a compact Hausdorff space and Φ is a commutative family of continuous self-maps of Q. Then there is an RBPM μ on the Borel subsets of Q that is Φ-invariant in the sense that for every $f \in C(Q)$ and $\varphi \in \Phi$,*

$$\int f \circ \varphi \, d\mu = \int f \, d\mu$$

or equivalently, for every Borel subset B of Q,

$$\mu(\varphi^{-1}(B)) = \mu(B).$$

Proof. Exercise.   □

Example 9.22. Let D denote the closed unit disc of $\mathbb{R}^2$, and Φ the restrictions to D of rotations of $\mathbb{R}^2$ about the origin. Then Φ is a commutative family of maps, each of which takes D continuously onto itself. Thus Corollary 9.21 guarantees for D an RBPM invariant under each member of Φ. In fact *two* such measures come immediately to mind: Lebesgue area measure on D and arc-length measure on ∂D, the unit circle (both measures normalized to have total mass one). Thus the uniqueness established above for Haar measure on compact abelian groups fails for the more general case of RBPMs invariant under commutative families of maps.

> *Exercise* 9.16. Show that there is an uncountable family of RBPM's on D invariant under the rotation group Φ defined above.

Remark 9.23 (Role of commutativity). In the arguments above, the hypothesis of commutativity imposed upon the family of maps Φ showed up only at the very end, where it legitimized our use of the Markov–Kakutani Theorem. In Chap. 12 we'll extend the Markov–Kakutani Theorem to families of maps that are "almost commutative," (e.g., to groups of maps that are *solvable*). Here the argument that proved Theorem 9.19 will go through *verbatim*, with commutativity replaced at the final stage by the new hypothesis on the family Φ of self-maps of S. As a corollary we'll obtain the existence of Haar measure for compact, solvable groups. To obtain Haar measure for *all* compact groups, however, will require a new fixed-point theorem; this we'll explore in Chap. 13.

9.7 Harmonic Analysis on Compact Abelian Groups

The existence of Haar measure for compact abelian groups allows us to generalize to that context the Fourier analysis that's so important for functions that are integrable on the unit circle $\mathbb{T}$. Recall that Haar measure μ on $\mathbb{T}$ is Lebesgue arc-length measure normalized to have unit mass. For $1 \le p \le \infty$, let's denote the (complex) Lebesgue space of $\mathbb{T}$ with respect to this measure by $L^p(\mathbb{T})$. Let $\gamma_n(\zeta) = \zeta^n$ for $\zeta \in \mathbb{T}$, and for $f \in L^1(\mathbb{T})$ and $n \in \mathbb{Z}$, define the *n-th Fourier coefficient of f* by

$$\hat{f}(n) = \int_{\mathbb{T}} f \gamma_n^{-1} d\mu = \frac{1}{2\pi} \int_0^{2\pi} f(e^{i\theta}) e^{-in\theta} d\theta = \langle f, \gamma^n \rangle,$$

where $d\mu(e^{i\theta}) = \frac{d\theta}{2\pi}$.

Now $L^2 = L^2(\mathbb{T})$ is a Hilbert space with inner product

$$\langle f, g \rangle = \int f \bar{g} d\mu \qquad (f, g \in L^2)$$

so $\hat{f} = \langle f, \gamma_n \rangle$ for $f \in L^2$ and $n \in \mathbb{Z}$. It's easy to check that the exponential functions $\{\gamma_n : n \in \mathbb{Z}\}$ form an orthonormal subset of L^2. The linear span $\mathscr{T}$ of this orthonormal set (the set of *trigonometric polynomials*) is a subalgebra of $C(\mathbb{T})$ that's closed under complex conjugation and separates points of $\mathbb{T}$ (a feat accomplished single-handedly by γ_1, which is the identity map on $\mathbb{T}$). Thus the Stone-Weierstrass Theorem (see, e.g., [101, Theorem 7.33, p. 165]) assures us that $\mathscr{T}$ is dense in the max-norm topology of $C(\mathbb{T})$, and since $C(\mathbb{T})$ is dense in L^2, and the L^2-topology is weaker than that of $C(\mathbb{T})$), we see that $\mathscr{T}$ is dense in L^2.

Conclusion: The exponential functions $\{\gamma_n : n \in \mathbb{Z}\}$ form an *orthonormal basis* for L^2, hence for every function f in that space we have

$$\|f\|_2^2 = \sum_{n \in \mathbb{Z}} |\hat{f}(n)|^2$$

from which it follows that

$$f = \sum_{n \in \mathbb{Z}} \langle f, \gamma_n \rangle \gamma_n = \sum_{n \in \mathbb{Z}} \hat{f}(n) \gamma_n \qquad (9.12)$$

with the series convergent in L^2 *unconditionally* in the sense that for every $\varepsilon > 0$ there exists a finite subset F_ε of $\mathbb{Z}$ such that

$$F \supset F_\varepsilon \implies \left\| \sum_{n \in F} \hat{f}(n) \gamma_n - f \right\| < \varepsilon.$$

The series in (9.12) is called the *Fourier series* of f; it represents a decomposition of that function into "frequencies" γ_n.

All this is standard, and forms the basis for the harmonic analysis of square-integrable functions on the unit circle. In order to generalize this theory to other compact abelian groups, we need the following observation:

Proposition 9.24. *The exponential functions $\{\gamma_n : n \in \mathbb{Z}\}$ are precisely the* continuous homomorphisms *of $\mathbb{T}$ into itself.*

One checks easily that each function γ_n is indeed a homomorphism of $\mathbb{T}$ (meaning: $\gamma_n(\zeta \eta^{-1}) = \gamma_n(\zeta)\gamma_n(\eta)^{-1}$ for each $\zeta, \eta \in \mathbb{T}$ and $n \in \mathbb{Z}$). The exercise set below shows that the γ_n's are the *only* continuous homomorphisms of $\mathbb{T}$.

Exercise 9.17. Suppose $\Gamma : \mathbb{R} \to \mathbb{T}$ is a continuous group homomorphism, where $\mathbb{R}$ has its additive structure. Thus $\Gamma(0) = 1$, $\Gamma(x+y) = \Gamma(x)\Gamma(y)$ and $\Gamma(-x) = \Gamma(x)^{-1}$ for each $x, y \in \mathbb{R}$.

(a)   Suppose in addition that Γ is *differentiable* at the origin. Show that Γ is differentiable at every $x \in \mathbb{R}$, with $\Gamma'(x) = \Gamma'(0)\Gamma(x)$. Conclude that $\Gamma(x) = e^{i\lambda x}$ for each $x \in \mathbb{R}$, where $\lambda := \Gamma'(0)$.

(b)   Not assuming the differentiability of Γ, show that there exists $\delta \in (0, 2\pi)$ for which $A := \int_0^\delta \Gamma(t) \, dt \neq 0$. Show that this implies $A\Gamma(x) = \int_{s=x}^{x+\delta} \Gamma(s) \, ds$. Conclude that Γ is differentiable on $\mathbb{R}$.

(c)   Show that if Γ is 2π-periodic on $\mathbb{R}$ then the constant λ of part (a) must be an integer.

(d)   Now use part (c) to finish the proof of Proposition 9.24.

Definition 9.25. A *character* of a topological group G is a continuous homomorphism $G \to \mathbb{T}$.

Notation. We'll use $\hat{G}$ to denote the set of characters of G. Thus $\hat{\mathbb{T}} = \{\gamma_n : n \in \mathbb{Z}\}$.

Exercise 9.18 (Dual Group). Show that $\hat{G}$, with pointwise multiplication, is a group. It's called the *dual group* of G.

Exercise 9.19 (Some Dual Groups). For $\lambda \in \mathbb{R}$ define the exponential function $\gamma_\lambda(x) := e^{i\lambda x}$ for $x \in \mathbb{R}$. Think of $\mathbb{R}$ as a group with its usual additive structure, and $\mathbb{Z}$ as a subgroup of $\mathbb{R}$. Show that:

(a) $\hat{\mathbb{R}}$ is group-isomorphic to $\mathbb{R}$ via the identification $\lambda \to \gamma_\lambda$, $\lambda \in \mathbb{R}$.

(b) $\hat{\mathbb{T}}$ is group-isomorphic to $\mathbb{Z}$ via the identification $n \to \gamma_n$, $n \in \mathbb{Z}$.

For a compact abelian group G it's common to denote by $L^2(G)$ the L^2-space defined for the Haar measure of G.

Exercise 9.20. If G is a compact abelian group then $\hat{G}$ is an orthonormal subset of $L^2(G)$.

Just as for the unit circle, $\hat{G}$ is an orthonormal *basis* for $L^2(G)$. As before, $\hat{G}$ is closed under complex conjugation (easy), and separates points of G (not easy: see, e.g., [100, Sect. 1.5.2, p. 24]), so its linear span—what we might think of as the collection of *trigonometric polynomials* on G—is, by Stone-Weierstrass, dense in $C(G)$, hence also in $L^2(G)$.

If $f \in L^2(G)$ and $\gamma \in G$, we define the *Fourier transform* $\hat{f} : \hat{G} \to \mathbb{C}$ by

$$\hat{f}(\gamma) := \int_G f\gamma^{-1}\,d\mu \qquad (\gamma \in \hat{G}),$$

where μ is Haar measure for G. Then, as in the circle case:

Proposition 9.26. *If G is a compact abelian group then for each $f \in L^2(G)$,*

$$\sum_{\gamma \in G} |\hat{f}(\gamma)|^2 = \|f\|_2 \qquad \text{and} \qquad \sum_{\gamma \in G} \hat{f}(\gamma)\gamma = f.$$

These formulae employ the same kind of "unordered summation" used for Fourier series on the circle group, with the term "Fourier series" once again denoting the character series representing f.

Remark. The characters of *non-commutative* groups never separate points. Indeed, if g and h are non-commuting elements of a group G, and γ is a character on G, then $\gamma(gh) = \gamma(g)\gamma(h) = \gamma(hg)$. It gets worse; Exercise 13.8 (p. 178) shows that even for compact groups it's possible for the only character to be the trivial one $\gamma \equiv 1$!

Notes

Commuting families of maps. Apropos to Example 9.4: A long-standing problem asked if every commuting family of continuous self-maps of the *closed unit interval* had to have a common fixed point. Counterexamples were published in 1969, independently by Boyce [15] and Hunecke [52]; their constructions are nontrivial.

Exercise on Borel sets. Thanks to Urysohn's Lemma the result of Exercise 9.3 remains true for normal (Hausdorff) topological spaces. However it is *not* true in general. Let X denote the space of ordinal numbers less than or equal to the first uncountable one Ω, taken in the interval topology. Then X is a Hausdorff space, but each continuous real-valued function on X is constant on some final segment $[\alpha, \Omega]$, α a countable ordinal. From this it follows that, although the singleton $\{\Omega\}$ is a Borel set (it is closed), it does not belong to $\mathscr{C}$, the smallest sigma algebra rendering each continuous real-valued function measurable. Thus $\mathscr{C}$ is strictly smaller than the sigma algebra of Borel subsets of X.

The Markov–Kakutani Theorem. The proof of Theorem 9.6 is Kakutani's proof from [56]. Markov earlier gave a proof [75] for locally convex spaces, using Tychonoff's extension [120] to that setting of the Schauder Fixed-Point Theorem.

The Tychonoff Product Theorem. For a proof that the Axiom of Choice implies the Tychonoff Product Theorem see [103, Theorem A2, pp. 392–393], or [55, Sect. 2.2, p. 11] for a proof based on "filters," or [25] for one based on "nets."

The Tychonoff Product Theorem does not require that the factors of the product be Hausdorff; in fact this "non-Hausdorff" version of the theorem is actually *equivalent* to the Axiom of Choice (see, e.g., [55, Sect. 2.6, p. 26, Problem 8]).

The Riesz Representation Theorem. The first result of this type appeared in a 1909 paper of Frigyes Riesz [97], who proved that every continuous linear functional on the Banach space $C([0,1])$ is represented by Stieltjes integration against a real-valued function on $[0,1]$ of bounded variation. The "positive" version we've been using above follows easily from this one.

Chapter 10
The Meaning of Means

FINITELY ADDITIVE MEASURES, EXTENSIONS, AND AMENABLE GROUPS

Overview. In the last chapter we used the Markov–Kakutani Theorem to produce for every compact abelian group G an "invariant mean": a mean in the algebraic dual space of $B(G)$ fixed by the adjoint of every translation map on $B(G)$. The Riesz Representation Theorem then provided a G-invariant, regular, Borel probability measure to represent this mean via integration on $C(G)$. Thus was born Haar measure for G.

In this chapter we will scale back the role of topology and observe that for each abelian group G this "Markov–Kakutani" mean easily provides a G-invariant, finitely additive "probability measure" on $\mathscr{P}(G)$, the collection of all subsets of G. We'll examine the significance of such set functions. In the compact case, might one of them extend Haar measure? Which non-abelian groups support such "measures"? Such questions will lead (next chapter) into the study of "paradoxical decompositions," most notably the celebrated Banach–Tarski Paradox.

Prerequisites. A little: measure theory, group theory, functional analysis.

10.1 Means and Finitely Additive Measures

We've previously attached to a set S the following cast of characters:

- $\mathscr{P}(S)$: The collection of all subsets of S.
- $B(S)$: The vector space of all bounded, real-valued functions on S.
- $B(S)^{\sharp}$: The *algebraic dual* of $B(S)$; all the linear functionals on $B(S)$.
- $\mathscr{M}(S)$: The *means* on $B(S)$; the collection of positive linear functionals Λ on $B(S)$ "normalized" so that $\Lambda(1) = 1$. We've noted that $\mathscr{M}(S)$ is a nonempty, convex subset of $B(S)^{\sharp}$ (Exercise 9.13).
- $\omega(S)$: The *weak-star topology* on $B(S)^{\sharp}$; the restriction to $B(S)^{\sharp}$ of the product topology of $\mathbb{R}^{B(S)}$. We've seen that $\mathscr{M}(S)$ is $\omega(S)$-compact (Corollary 9.17).

© Springer International Publishing Switzerland 2016
J.H. Shapiro, *A Fixed-Point Farrago*, Universitext,
DOI 10.1007/978-3-319-27978-7_10

"Measures" from Means. Each mean Λ on $B(S)$ naturally defines a function μ on the collection $\mathscr{P}(S)$ of all subsets of S:

$$\mu(E) = \Lambda(\chi_E) \qquad (E \in \mathscr{P}(S)), \tag{10.1}$$

where χ_E denotes the characteristic function of E ($\equiv 1$ on E and $\equiv 0$ off E).

Exercise 10.1. For μ as defined above, show that:

(a) $\mu(S) = 1$.

(b) μ is *monotone*: $E \subset F \subset S \implies \mu(E) \leq \mu(F)$.

(c) $\mu(E) \leq 1$ for every $E \subset S$.

The linearity of Λ translates into *finite additivity* for μ: if $\{E_1, E_2, \ldots E_n\}$ is a finite, pairwise disjoint collection of subsets of S then $\chi_{\cup_k E_k} = \sum_k \chi_{E_k}$, so

$$\mu\left(\bigcup_k E_k\right) = \Lambda(\chi_{\cup_k E_k}) = \Lambda\left(\sum_k \chi_{E_k}\right) = \sum_k \Lambda(\chi_{E_k}) = \sum_k \mu(E_k).$$

Definition 10.1. A *finitely additive probability measure* on $\mathscr{P}(S)$ is a finitely additive function $\mu : \mathscr{P}(S) \to [0, 1]$ with $\mu(S) = 1$.

In this terminology the argument above established:

Proposition 10.2. *Each mean Λ on $B(S)$ induces via Eq. (10.1) a finitely additive probability measure μ on $\mathscr{P}(S)$.*

The exercise below shows that conversely each finitely additive probability measure μ on $\mathscr{P}(S)$ gives rise to a mean on $B(S)$, created as a sort of "Riemann integral."

Exercise 10.2 (Means from "Measures"). Let $\mathscr{S}(S)$ denote the collection of "simple functions" on S, i.e., the functions $f : S \to \mathbb{R}$ that take on only finitely many values.

Given a finitely additive probability measure μ on $\mathscr{P}(S)$ and a simple function f on S with distinct values $\{a_j\}_{j=1}^n$, let $E_j = f^{-1}(a_j)$ and define $\Lambda(f) := \sum_{j=1}^n a_j \mu(E_j)$.

(a)   Check that $\mathscr{S}$ is a vector space on which the functional Λ is positive and linear, and that Λ obeys the inequality promised for means by Exercise 9.14.

(b)   Show that Λ has a unique extension to a mean on $B(S)$ [Hint: Show that Λ is continuous if $\mathscr{S}$ is given the "sup-norm" $\|\cdot\|$ defined on $B(S)$ by Eq. (9.10)].

Invariant Means. Theorem 9.19 told us that if Φ is a commutative family of self-maps of the set S, then $B(S)$ has a mean Λ that is Φ-*invariant* in the sense that $C_\varphi^\sharp \Lambda = \Lambda$ for every $\varphi \in \Phi$, where $C_\varphi : B(S) \to B(S)$ is the composition operator $f \to f \circ \varphi$ defined on p. 114. The finitely additive probability measure μ that Λ induces on $\mathscr{P}(S)$ via Eq. (10.1) inherits this Φ-invariance:

$$\mu(\varphi^{-1}(E)) = \Lambda(\chi_{\varphi^{-1}(E)}) = \Lambda(\chi_E \circ \varphi)) = (C_\varphi^\sharp \Lambda)(\chi_E) = \Lambda(\chi_E) = \mu(E)$$

for every $E \subset S$ and $\varphi \in \Phi$. In summary:

Theorem 10.3. *If* Φ *is a commutative family of self-maps of a set* S *then there is a finitely additive probability measure* μ *on* $\mathscr{P}(S)$ *that is* Φ-invariant *in the sense that* $\mu(\varphi^{-1}(E)) = \mu(E)$ *for every* $E \in \mathscr{P}(S)$ *and every* $\varphi \in \Phi$.

Corollary 10.4. *If* G *is a commutative group then there exists a finitely additive probability measure* μ *on* $\mathscr{P}(G)$ *that is* G-invariant *in the sense that* $\mu(gE) = \mu(E)$ *for each* $g \in G$ *and* $E \in \mathscr{P}(G)$.

Proof. Apply Theorem 10.3 with $S = G$ and Φ the collection of "translation maps" $x \to g^{-1}x$ for g and x in G. □

Suppose in Theorem 10.3 we take Φ to be the group of rotations of $\mathbb{R}^2$ about the origin and S to be either of the following subsets of $\mathbb{R}^2$: the closed unit disc $\mathbb{B}^2$, or its boundary $\mathbb{T}$, the unit circle. In either case Φ is a commutative family of self-maps of S, hence Theorem 10.3 yields:

Corollary 10.5. *Both* $\mathscr{P}(\mathbb{B}^2)$ *and* $\mathscr{P}(\mathbb{T})$ *support a rotation-invariant, finitely additive probability measure.*

The question arises for either case: can such a finitely additive, rotation-invariant probability measure be chosen to agree, on Borel sets, with normalized Lebesgue measure. Similarly, for every compact abelian group, must the invariant measure promised by Corollary 10.4 agree on Borel sets with Haar measure? We'll study this matter of invariant extension in the next section. Not surprisingly, it will involve the Hahn–Banach Theorem.

10.2 Extending Haar Measure

Suppose G is a compact abelian group. We now know that G has both:

- a G-invariant regular probability measure μ on its Borel sets (Haar measure: Corollary 9.20), and
- a G-invariant *finitely additive* probability measure ν on its collection $\mathscr{P}(G)$ of *all* subsets (Theorem 10.3).

Since ν arose from a G-invariant mean on $B(G)$, and μ (via the Riesz Representation Theorem) from the restriction of that mean to $C(G)$, one might suspect that ν extends Haar measure from the Borel subsets of G to all of $\mathscr{P}(G)$, i.e., that the restriction of ν to the Borel subsets of G is μ. Surprisingly, this need not be the case; Banach proved in 1923 that it fails for the circle group $\mathbb{T}$.

> There exists a rotation-invariant, finitely additive probability measure on $\mathscr{P}(\mathbb{T})$ that does not extend Haar measure [6, Théorème 20].

We'll see later that for a compact group: there can be *at most one* Haar measure (Chap. 12), and that *there always is* a Haar measure (Chap. 13). In particular, for the unit circle $\mathbb{T}$, normalized Lebesgue measure is the unique rotation-invariant regular Borel measure. Thus Banach's result can be restated:

There exists a rotation-invariant, finitely additive probability measure on $\mathscr{P}(\mathbb{T})$ whose restriction to the Borel subsets of $\mathbb{T}$ is not countably additive.

In view of Banach's result, it makes sense to ask if Haar measure on G can be extended to a finitely additive G-invariant probability measure on $\mathscr{P}(G)$. Thanks to the Markov–Kakutani Theorem the answer is affirmative, with the desired extension of Haar measure following from an "invariant" version of the Hahn–Banach Theorem. First recall the usual version:

The Hahn–Banach Theorem. *Suppose V is a vector space over the real field and $p\colon V \to \mathbb{R}$ is a gauge function on V, i.e.,*

(a)   *$p(u+v) \le p(u) + p(v)$ for all $u, v \in V$, and*
(b)   *$p(av) = ap(v)$ for every $a \in \mathbb{R}$ with $a \ge 0$ and every $v \in V$.*

Suppose W is a linear subspace of V and Λ is a linear functional on W for which $\Lambda(w) \le p(w)$ for all $w \in W$. Then Λ has a linear extension $\tilde{\Lambda}$ to V such that

$$\tilde{\Lambda}(v) \le p(v) \quad \text{for all } v \in V.$$

Now consider that problem of extending Haar measure μ from the Borel subsets of a compact abelian group G to a finitely additive measure ν on $\mathscr{P}(G)$. The measure μ induces, via integration, a G-invariant linear functional Λ on $C(G)$, where we now view $C(G)$ as a linear subspace of $B(G)$. In order to make the desired extension of μ it will be enough to extend Λ to a G-invariant mean on $B(G)$. This will be accomplished by:

Theorem 10.6 (The "Invariant" Hahn–Banach Theorem). *Suppose V is a vector space and $\mathscr{G}$ is a commutative family of linear transformations $V \to V$. Suppose W is a linear subspace of V that is taken into itself by every transformation in $\mathscr{G}$, and that p is a gauge function on V that is "$\mathscr{G}$-subinvariant" in the sense that*

$$p(\gamma(v)) \le p(v) \quad \text{for every } v \in V \text{ and } \gamma \in \mathscr{G}.$$

If Λ is a $\mathscr{G}$-invariant linear functional on W that is dominated by p, i.e.,

$$\Lambda \circ \gamma = \Lambda \text{ for all } \gamma \in \mathscr{G} \quad \text{and} \quad \Lambda(v) \le p(v) \text{ for all } v \in W,$$

then Λ has a $\mathscr{G}$-invariant linear extension to V that is dominated on V by p.

Proof. Endow $V^\sharp$, the algebraic dual of V, with the weak-star topology ω induced on it by V. Let $\mathscr{K}$ be the collection of all linear extensions of Λ to V that are dominated on V by p. Clearly $\mathscr{K}$ is a convex subset of $V^\sharp$. By the (usual) Hahn–Banach Theorem, $\mathscr{K}$ is nonempty.

Claim: $\mathscr{K}$ *is weak-star compact in $V^\sharp$.*

Proof of Claim. By Corollary 9.15 we need only show that $\mathscr{K}$ is pointwise bounded on V and weak-star closed in $V^\sharp$. If $\tilde{\Lambda} \in \mathscr{K}$ then for every $x \in V$ we have, in addition to the defining property $\tilde{\Lambda}(x) \le p(x)$, also $-\tilde{\Lambda}(x) = \tilde{\Lambda}(-x) \le p(-x)$. Thus

$$-p(-x) \leq \tilde{\Lambda}(x) \leq p(x) \qquad (x \in V, \ \tilde{\Lambda} \in \mathscr{K}) \tag{10.2}$$

so $\mathscr{K}$ is pointwise bounded on V.

To see that $\mathscr{K}$ is weak-star closed in $V^{\sharp}$, suppose $\Lambda_0 \in V^{\sharp}$ is a weak-star limit point of $\mathscr{K}$. We wish to show that $\Lambda_0 \in \mathscr{K}$, i.e., that Λ_0 is an extension of Λ from W to V that's dominated by p. To see that Λ_0 extends V, fix $w \in W$ and $\varepsilon > 0$. Then the weak-star basic neighborhood

$$N(\Lambda_0, \{w\}, \varepsilon) = \{\Lambda \in V^{\sharp} : |\Lambda(w) - \Lambda_0(w)| < \varepsilon\}$$

contains a linear functional $\Lambda_1 \in \mathscr{K}$. Thus $|\Lambda_0(w) - \Lambda(w)| = |\Lambda_0(w) - \Lambda_1(w)| < \varepsilon$, whereupon $\Lambda_0(w) = \Lambda(w)$ because ε is an arbitrary positive number; hence Λ_0 is an extension of Λ to V.

Similarly, fix $v \in V$ and $\varepsilon > 0$. Choose $\Lambda_2 \in \mathscr{K} \cap N(\Lambda_0, \{v\}, \varepsilon)$. Then $|\Lambda_0(v) - \Lambda_2(v)| < \varepsilon$, so $\Lambda_0(v) < \Lambda_2(v) + \varepsilon \leq p(v) + \varepsilon$, hence $\Lambda_0(v) \leq p(v)$, once again by the arbitrariness of ε. This completes the proof of the Claim.

Finally, since each $\gamma \in \mathscr{G}$ is a linear map $V \to V$, it has an *adjoint* $\gamma^{\sharp} : V^{\sharp} \to V^{\sharp}$. Let $\mathscr{G}^{\sharp} := \{\gamma^{\sharp} : \gamma \in \mathscr{G}\}$. One checks easily that $\mathscr{G}^{\sharp}$ is a commutative family of linear maps on $V^{\sharp}$, each of which, thanks to the $\mathscr{G}$-subinvariance of the gauge function p, takes $\mathscr{K}$ into itself. By Theorem 9.18 each map $\gamma^{\sharp}$ is ω-continuous, hence the triple $(V^{\sharp}, \mathscr{K}, \mathscr{G}^{\sharp})$, with $V^{\sharp}$ carrying its weak-star topology, satisfies the hypotheses of the Markov–Kakutani theorem.

Conclusion: There exists $\tilde{\Lambda} \in K$ fixed by $\mathscr{G}^{\sharp}$, i.e.,

$$\tilde{\Lambda} \circ \gamma = \gamma^{\sharp}(\tilde{\Lambda}) = \tilde{\Lambda} \quad \text{for every} \quad \gamma \in \mathscr{G}.$$

This functional $\tilde{\Lambda}$ is the desired $\mathscr{G}$-invariant extension of our original one Λ. $\qquad \square$

Here, stated in generality, is our application to extension of invariant measures.

Corollary 10.7. *Let S be a compact Hausdorff space upon which acts a commutative family Φ of continuous mappings. Suppose μ is a (countably additive) Φ-invariant probability measure on the Borel subsets of S. Then μ extends to a Φ-invariant, finitely additive probability measure on $\mathscr{P}(S)$.*

Proof. Let Λ be the positive linear functional defined on $C(S)$ by integration against μ. By the invariance of μ and the change-of-variable formula of measure theory, Λ is invariant for each of the composition operators C_{φ} on $C(S)$ in the sense that $\Lambda \circ C_{\varphi} = \Lambda$ for each $\varphi \in \Phi$. Define the gauge function p on $B(S)$ by

$$p(f) = \|f\| = \sup_{s \in S} f(s) \qquad (f \in B(S)).$$

Clearly: p is C_{Φ}-invariant (in the sense that $p \circ C_{\varphi} = p$ for every $\varphi \in \Phi$), and $\Lambda \leq p$ on $C(S)$.

The Invariant Hahn–Banach Theorem now supplies an extension of Λ to a linear functional $\tilde{\Lambda}$ on $B(S)$ that's also dominated by p, and is invariant for each mapping C_{φ} for $\varphi \in \Phi$. Upon applying inequality (10.2) to our gauge function p, we see that

$$\inf_{s\in S} f(s) \leq \tilde{\Lambda}(f) \leq \sup_{s\in S} f(s) \qquad (f\in B(S)),$$

so if $f \geq 0$ on S then $\tilde{\Lambda}(f) \geq 0$, i.e., $\tilde{\Lambda}$ is a positive linear functional on $B(S)$. Since $\tilde{\Lambda}(1) = \Lambda(1) = 1$, the functional $\tilde{\Lambda}$ is a mean on $B(S)$. The desired extension $\tilde{\mu}$ of μ to $\mathscr{P}(S)$ now emerges from Eq. (10.1) with $\tilde{\Lambda}$ in place of Λ, the Φ-invariance of $\tilde{\mu}$ following from the C_{Φ}-invariance of $\tilde{\Lambda}$. $\qquad\square$

For our original problem of extending Haar measure on a compact abelian group G, we take in Corollary 10.7: $S = G$ and $\Phi =$ the set of translation maps $x \to g^{-1}x$ for g and x in G. The result:

Corollary 10.8. *For each compact abelian group G, Haar measure has an extension to a finitely additive G-invariant measure on $\mathscr{P}(G)$.*

Since the group of rotations of $\mathbb{R}^2$ about the origin is abelian, Corollary 10.7 yields

Corollary 10.9. *There is a rotation-invariant, finitely additive probability measure on the closed unit disc of $\mathbb{R}^2$ that extends Lebesgue area measure from the Borel sets to all subsets. The unit circle supports a similar extension of normalized arc-length measure.*

Our final application of the invariant Hahn–Banach theorem involves the creation of a notion of "limit" for every bounded real sequence. We'll use the notation ℓ^{∞} for the space of all such sequences.

Corollary 10.10 (Banach limits). *There exists a positive, translation-invariant linear functional Λ on ℓ^{∞} such that*

$$\liminf_{n\to\infty} f(n) \leq \Lambda(f) \leq \limsup_{n\to\infty} f(n) \qquad (f\in \ell^{\infty}).$$

Proof. Let c denote the space of real sequences $f: \mathbb{N} \to \mathbb{R}$ for which $\lambda(f) = \lim_{n\to\infty} f(n)$ exists (in $\mathbb{R}$). For $f \in \ell^{\infty}$ let

$$p(f) = \limsup_{n\to\infty} f(n).$$

Then p is a gauge function on ℓ^{∞}, and $\lambda \leq p$ on c. For $k \in \mathbb{N}$ define the "translation map" T_k on ℓ^{∞} by

$$T_k f(n) = f(n+k) \qquad (f\in \ell^{\infty}, n\in \mathbb{N}).$$

Thus $\mathscr{T} = \{T_k : k \in \mathbb{N}\}$ is a commutative family of linear transformations $\ell^{\infty} \to \ell^{\infty}$ for each of which: the subspace c is taken into itself, and both λ and p are invariant. Thus the Invariant Hahn–Banach Theorem applies and produces a $\mathscr{T}$-invariant extension Λ of λ to ℓ^{∞} with $\Lambda \leq p$ on ℓ^{∞}. By inequality (10.2):

$$\liminf_{n\to\infty} f(n) = -p(-f) \leq \Lambda(f) \leq p(f) = \limsup_{n\to\infty} f(n) \qquad (f\in \ell^{\infty}). \qquad\square$$

The functional Λ produced above is called a *Banach limit*; the usual notation is $\Lambda(f) := \mathrm{LIM}_{n\to\infty} f(n)$.

> *Exercise* 10.3. Each Banach limit defines a translation-invariant finitely additive probability measure μ on $\mathscr{P}(\mathbb{N})$ by: $\mu(E) := \mathrm{LIM}_n\, \chi_E(n)$ for $E \subset \mathbb{N}$.
>
> (a)   Show that $\mu(\{n\}) = 0$ for every $n \in \mathbb{N}$. Conclude that μ is *not* countably additive.
>
> (b)   For n_0 and k in $\mathbb{N}$, let E denote the arithmetic progression $\{n_0 + kn : n \in \mathbb{N} \cup \{0\}\}$. What is $\mu(E)$?
>
> (c)   Is there an infinite subset E of $\mathbb{N}$ with $\mu(E) = 0$?

This exercise points the "Jekyll and Hyde" character possessed by an infinite dimensional vector space's algebraic dual. On one hand, the algebraic dual is easy to define and work with (e.g., no worries about continuity). On the other hand, thanks to the Axiom of Choice it has bizarre inhabitants (e.g., Banach limits).

> *Exercise* 10.4 ("Banach limits" for $\mathbb{Z}$ and $\mathbb{R}$). Show that analogues of "Banach Limit" exist for the additive groups of both the integers and the real line.

10.3 Amenable Groups

Thanks to Corollary 10.4 we know that every abelian group G possesses an *invariant mean*, i.e., a positive linear functional Λ on $B(G)$ that takes value 1 on the constant function 1 and is fixed by the adjoint of every operator of translation by a group element. We've noted that such a mean gives rise to a finitely additive probability measure μ on $\mathscr{P}(G)$ that's *G-invariant* in the sense that $\mu(gE) = \mu(E)$ for each $g \in G$ and $E \in \mathscr{P}(G)$.

Definition 10.11 (Amenable group). To say a group G is *amenable* means that there is a G-invariant, finitely additive probability measure on $\mathscr{P}(G)$, i.e., there is a G-invariant mean on $B(G)$.

Thus every abelian group is amenable. What about the non-abelian ones? Once we venture into the realm of non-commutativity there arises the spectre of "left vs. right." For non-abelian groups the sort of invariance we've been considering should more accurately be called "left-invariance."

Question. Are there separate notions of "right-" and "left-" amenable?

We'll see later on (Sect. 12.6) that once a group has a left-invariant mean, then it also has a right-invariant one, and even a "bi-invariant" one. So there are not separate concepts of "left-amenable" and "right-amenable"; it's all just "amenable."

> *Exercise* 10.5. Show that every finite group is amenable.

It turns out that *not every* group is amenable. Here's an example, whose apparent simplicity belies its importance:

The Free Group F_2 on Two Generators. The elements of F_2 are "reduced words" of the form $x_1 x_2 \cdots x_n$ for $n \in \mathbb{N}$ where each x_j comes from the set of symbols $\{a, a^{-1}, b, b^{-1}\}$, subject only to the restriction that no symbol occurs next to its "inverse." Multiplication in F_2 is defined to be concatenation of words, followed by "reduction," e.g., $aba^{-1} \cdot abba = abbba$. Upon allowing the "empty word" e to belong to F_2 we obtain a group.

> *Caveat:* To render the group operation of F_2 "well-defined" it must be shown that the same reduced word results no matter how this reduction is performed. This is not completely trivial (see, e.g., [73, Theorem 1.2, pp. 134–5]). In the next chapter we'll resolve this matter differently by realizing F_2 as a group of rotations of $\mathbb{R}^3$.

> *Exercise* 10.6. Convince yourself that (modulo the above *caveat*) F_2 is a group, that it's not abelian, and that it can be visualized as the fundamental group of a figure-eight.

Theorem 10.12. *F_2 is not amenable.*

Proof. For $x \in \{a, a^{-1}, b, b^{-1}\}$ let $W(x)$ denote the set of reduced words that begin with x. For example, a and $ab^{-1}abb$ belong to $W(a)$, while b and $a^{-1}baab^{-1}$ do not. Thus the sets $W(a), W(a^{-1}), W(b)$, and $W(b^{-1})$ form a pairwise disjoint family of sets in F_2 whose union is $F_2 \setminus \{e\}$. Note that $aW(a^{-1})$ is the set of reduced words in F_2 that don't begin with a, so F_2 is the disjoint union of $W(a)$ and $aW(a^{-1})$; similarly it's also the disjoint union of $W(b)$ and $bW(b^{-1})$.

Now suppose for the sake of contradiction that μ is a finitely additive probability measure on $\mathscr{P}(F_2)$ that is F_2-invariant. Then, upon using disjointness in the third line below and the invariance of μ in the fourth, we obtain

$$
\begin{aligned}
1 &\geq \mu(F_2 \setminus \{e\}) \\[4pt]
&= \mu\big(W(a) \cup W(a^{-1}) \cup W(b) \cup W(b^{-1})\big) \\[4pt]
&= \mu\big(W(a)\big) + \mu\big(W(a^{-1})\big) + \mu\big(W(b)\big) + \mu\big(W(b^{-1})\big) \\[4pt]
&= \mu\big(W(a)\big) + \mu\big(aW(a^{-1})\big) + \mu\big(W(b)\big) + \mu\big(bW(b^{-1})\big) \\[4pt]
&= \mu\big(\underbrace{W(a) \cup aW(a^{-1})}_{=\,F_2}\big) + \mu\big(\underbrace{W(b) \cup bW(b^{-1})}_{=\,F_2}\big) \\[4pt]
&= 1 + 1 = 2,
\end{aligned}
$$

i.e., $1 \geq 2$: a contradiction.   □

The question of which groups are amenable is a profound one. We'll see in chapters to come that every solvable group is amenable, but that some compact groups are not. Amenability is intimately connected with the phenomenon of *paradoxicality* which we'll take up in the next chapter; the free group F_2 will play a crucial role.

Notes

The Hahn–Banach Theorem. See, for example, [9, Chap. II, pp. 27–29], [60, Theorem 3.4, p. 21] or [103, Theorem 3.2, pp. 57–58] for the "non-invariant" version. Banach proved a precursor of the Hahn–Banach Theorem in the course of showing that there's a rotation-invariant mean on $B(\mathbb{T})$ whose resulting finitely additive probability measure on $\mathscr{P}(\mathbb{T})$ is *not* an extension of arc-length measure on the Lebesgue measurable subsets of $\mathbb{T}$ [6, Theorem 19–20]. Banach's result answered the one dimensional case of a more general problem posed by one of his former professors, Stanisław Ruziewicz.

The Ruziewicz Problem. This problem asks if Lebesgue surface measure on the unit sphere of $\mathbb{R}^{n+1}$ is the unique (up to multiplication by a positive constant) finitely additive, isometry-invariant measure on the Lebesgue measurable subsets of the sphere. The result of Banach mentioned above shows that the answer is "no" for $n = 1$. For $n > 1$ the problem remained open until the 1980s, when the answer was shown to be "yes" by Drinfeld [32] for $n = 2$ and 3, and for $n > 3$ independently by Margulis [74] and Sullivan [114].

The Invariant Hahn–Banach Theorem. This is due to Agnew and Morse [1]; the proof given here is taken from [37, Sects. 3.3 and 3.4].

Banach limits. The result here is due (with a different proof) to Banach [9, Chap. II, p. 34], who also noted the connection with finitely additive probability measures on the subsets of the positive integers [9, Remarques, Sect. 3, p. 231].

Amenable groups. In the 1920s von Neumann [88] initiated the study of groups G for which $\mathscr{P}(G)$ supports invariant finitely additive probability measures. He called such groups "measurable." The currently preferred term "amenable" was coined by M.M. Day in the late 1950s [28], reputedly as something of a pun on the term "mean" (see [104, p. 34], for example).

Chapter 11
Paradoxical Decompositions

SET-THEORETIC PARADOXES OF HAUSDORFF AND BANACH–TARSKI

Overview. In Chap. 10 we used the fixed-point theorem of Markov and Kakutani to show that every abelian group G is "amenable" in the sense that there is a G-invariant mean on the vector space $B(G)$ of bounded, real-valued functions on G. We observed that existence of such a "mean" is equivalent to existence of a finitely additive probability "measure" on $\mathscr{P}(G)$, the algebra of all subsets of G, and we asked if *every* group turns out to be amenable. We showed that the free group F_2 on two generators is *not* amenable by finding within F_2 four pairwise disjoint subsets that could be reassembled, using only group motions, into *two* copies of F_2.

Now we'll see how this "paradoxical" property of F_2, along with the Axiom of Choice, leads to astonishing results in set theory, most notably the famous Banach–Tarski Paradox, often popularly phrased as: *Each (three dimensional) ball can be partitioned into a finite collection of subsets which can then be reassembled, using only rigid motions, into* two *copies of itself.* Even more striking: given two bounded subsets of $\mathbb{R}^3$ with nonvoid interior, each can be partitioned into a finite collection of subsets that can be rigidly reassembled into the other. For this result the fixed-point theorem of Knaster and Tarski (Theorem 1.2) makes another appearance, this time to prove a far-reaching generalization of the Schröder–Bernstein Theorem.

Prerequisites. Elementary properties of: sets, groups, matrices.

11.1 Paradoxical Sets

To establish non-amenability for the free group F_2 on the two generators a and b (Theorem 10.12) we observed that its pairwise disjoint family of subsets $\mathscr{W} = \{W(a), W(a^{-1}), W(b), W(b^{-1})\}$ could be "F_2-reassembled" into *two* copies of F_2 in the sense that

$$F_2 = W(a) \uplus aW(a^{-1}) = W(b) \uplus bW(b^{-1}),$$

© Springer International Publishing Switzerland 2016
J.H. Shapiro, *A Fixed-Point Farrago*, Universitext,
DOI 10.1007/978-3-319-27978-7_11

where the symbol "$\uplus$" denotes "union of pairwise disjoint sets," and $W(x)$ denotes the collection of reduced words in the generators and their inverses that begin with the letter x. Although the family of subsets $\mathcal{W}$ does not exhaust all of F_2 (its union omits the empty word, a.k.a. the identity element of F_2), it can be easily modified to give a more symmetric statement.

Proposition 11.1. *There exists a pairwise disjoint family* $\{E_1, E_2, E_3, E_4\}$ *of subsets of* F_2 *such that*

$$F_2 = E_1 \uplus \dots \uplus E_4 = E_1 \uplus aE_2 = E_3 \uplus bE_4. \tag{11.1}$$

Proof. The provisions of (11.1) are fulfilled with $E_1 = W(a) \setminus \{a, a^2, a^3, \dots\}$, $E_2 = W(a^{-1}) \uplus \{e, a, a^2, a^3, \dots\}$, $E_3 = W(b)$, and $E_4 = W(b^{-1})$. $\quad\square$

This "paradoxical" nature of F_2 has far-reaching consequences. To see how it works, assume that X is an arbitrary set and G a group of self-maps of X. Thus G is a family of self-maps of X that is closed under composition, contains the identity map on X, and contains the (compositional) inverse of each of its members. In particular, each $g \in G$ is a *bijection* of X: a one-to-one mapping taking X onto itself. To say that a set is *partitioned* by a family of subsets means that the subsets of the family are nonempty, pairwise disjoint, and that their union is the whole set. We'll call such a subset family a *partition* of the ambient set.

Definition 11.2 (Paradoxical set). To say that a subset E of X is *G-paradoxical* means that there exist:

(a)   A partition of E into finitely many subsets $\{E_1, E_2, \dots, E_n\}$,
(b)   A collection $\{g_1, g_2, \dots g_n\}$ of elements of G, and
(c)   An integer $1 \le m < n$, such that each family $\{g_j E_j\}_1^m$ and $\{g_j E_j\}_{m+1}^n$ is a partition of E.

Thus "E is G-paradoxical" means that E has a partition whose members can be disjointly reassembled, via transformations in G, into *two* copies of E.

If the group G is understood, we'll abbreviate "G-paradoxical" to just "paradoxical." To say that G itself is paradoxical means that it's paradoxical with respect to the group of left-translation mappings $x \to gx$ $(x, g \in G)$ it induces upon itself. Thus Proposition 11.1 shows that F_2 is paradoxical, with $n = 4$ and $m = 2$ in Definition 11.2. We closed the previous chapter by showing that F_2 is not amenable. In fact this is true of *every* paradoxical group, as the following exercise shows.

> *Exercise* 11.1. Show that: if a set X is paradoxical with respect to a group G of its self-mappings, then $\mathcal{P}(X)$ supports no G-invariant finitely additive probability measure. *Corollary.* No paradoxical group is amenable.

Corollary 11.3. *Neither the closed unit disc* Δ *of* $\mathbb{R}^2$ *nor the unit circle* $\mathbb{T}$ *is paradoxical with respect to the group of rotations about the origin.*

Proof. This follows immediately from Exercise 11.1 above, thanks to Corollary 10.9, which establishes the existence of rotation-invariant finitely additive probability measures on $\mathcal{P}(\Delta)$ and $\mathcal{P}(\mathbb{T})$. $\quad\square$

Exercise 11.2. In Definition 11.2 let's call the set E *paradoxical using n pieces*, or more succinctly: *n-paradoxical*. For example, the free group F_2 is 4-paradoxical.

(a)   Show that the ambient set X itself cannot be n-paradoxical for $n < 4$.

(b)   Show that a *subset* of X can be n-paradoxical for $n < 4$. (Suggestion: Show that with respect to the bijection group of $\mathbb{Z}$, the subset of natural numbers is 2-paradoxical.)

The next result asserts that paradoxicality can often be transferred from a group to a set upon which that group acts; it is the key to all that follows.

Definition 11.4. To say such a group G of self-maps of a set X is *fixed-point free* on a subset E of X means that no element of G, other than the identity map, can fix a point of E.

Theorem 11.5 (The Transference Theorem). *Suppose X is a set and G a fixed-point free group of self-maps of X. If G is paradoxical, then X is G-paradoxical.*

Proof. We're given: a "replicator family" $\{E_j\}_1^n$ that partitions G, a corresponding family $\{g_j\}_1^n$ of elements of G and an integer m with $1 \leq m < n$ such that each "replicant family" $\{g_j E_j\}_{j=1}^m$ and $\{g_j E_j\}_{j=m+1}^n$ also partitions G. We want to show that this situation can be "lifted" to X.

For $x \in X$ the subset $Gx = \{gx : g \in G\}$ is called the G-*orbit of* x. It's easy to check (exercise) that: *The G-orbits partition X.* Consequence: we have $X = \uplus_{m \in M} Gm$ where $M \subset X$ is a "choice set" consisting of one element chosen from each G-orbit.[1] For $g \in G$ let's call the set $gM = \{gm : m \in M\}$ the "co-orbit" of g. The key to transference is the following:

Claim: The co-orbits partition X, i.e., $X = \uplus \{gM : g \in G\}$.

Proof of Claim. Observe first that *the co-orbits exhaust* X (proof: $\cup_{g \in G} gM = GM = X$). Thus we need only show that the co-orbits are pairwise disjoint. To this end suppose g and h belong to G and $gM \cap hM \neq \emptyset$. Then there exist points $m_1, m_2 \in M$ such that $gm_1 = hm_2$, so $h^{-1} gm_1 = m_2$, hence m_2 belongs to the G-orbit of m_1. By the definition of our choice set M we must therefore have $m_1 = m_2$, which provides a fixed point for the map $h^{-1} g \in G$. Since G is fixed-point free on X this forces $h^{-1} g$ to be the identity map on X, so $h = g$ and therefore $gM = hM$. Thus, given g and h in M with $g \neq h$, the co-orbits gM and hM must be disjoint, as desired.

For $A \subset G$ let $A^* = AM = \{a(m) : m \in M, a \in A\}$. Then thanks to the Claim:

If the family of sets $\{A_j\}_1^n$ partitions G then $\{A_j^*\}_1^n$ partitions X.

Figure 11.1 illustrates the situation.

---

[1] *Warning:* In general we need the Axiom of Choice (Appendix E.3, p. 209) to do this.

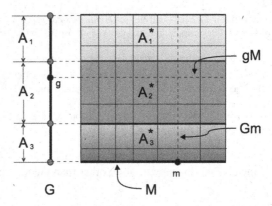

Fig. 11.1 $G = A_1 \uplus A_2 \uplus A_3 \implies X = A_1^* \uplus A_2^* \uplus A_3^*$

Thus our replicator partition $\{E_j\}_1^n$ of G can be transferred to a partition $\{E_j^*\}_1^n$ of X. Similarly the replicant partitions $\{g_j E_j\}_1^m$ and $\{g_j E_j\}_{m+1}^n$ of G transfer to replicant partitions $\{g_j E_j^*\}_1^m$ and $\{g_j E_j^*\}_{m+1}^n$ of X, establishing the G-paradoxicality of X.   □

Corollary 11.6. *Each group with a paradoxical subgroup is itself paradoxical.*

Proof. Every subgroup acts freely, by group multiplication, on its parent group. Thus by Theorem 11.5, if the subgroup is paradoxical then so is its parent.   □

Exercise 11.3. Show that when G acts freely on X, the family of co-orbits is *transverse* to the family of orbits, i.e., the intersection of each co-orbit with an orbit is a singleton.

Exercise 11.4 (Converse to Theorem 11.5). Suppose G is a group of self-maps of a set X. Show that if X is G-paradoxical, then G is paradoxical. (For this one it's not necessary that G act freely on X.) *Suggestion.* Suppose $\{E_j^*\}_1^n$, $\{g_j\}_1^n$, and $1 \leq m < n$ "witness" the G-paradoxicality of X. Fix $x \in X$ and define $E_j = \{g \in G : gx \in E_j^*\}$. Show that the E_j's, $g_j's$, and m witness paradoxicality for G.

Exercise 11.5. For a group G of self-maps of a set X, let C denote the set of points of X, each of which is fixed by some non-identity element of G. Show each map in G takes C, and therefore $X \backslash C$, onto itself. Thus G is a set of self-maps of $X \backslash C$ that is fixed-point free on that set.

11.2 The Hausdorff Paradox

In this section we'll work on the unit sphere S^2 of $\mathbb{R}^3$: the set of points of three dimensional euclidean space that lie at distance 1 from the origin. Let $\mathscr{R}$ denote the group of rotations of $\mathbb{R}^3$ about the origin. For the rest of this chapter we'll treat the notion of "three dimensional rotation" intuitively, taking for granted that each rotation has a "center" through which passes an "axis," every point of which it fixes, and that $\mathcal{R}$ is a group under composition—a group that acts on S^2. All these facts are established in Appendix D (Theorem D.7), where it's shown that $\mathcal{R}$ is isomorphic to the group SO(3) of 3×3 orthogonal matrices with determinant 1.

Theorem 11.7 (The Hausdorff Paradox, c. 1914). *$S^2 \setminus C$ is $\mathcal{R}$-paradoxical for some countable subset C of S^2.*

This result follows quickly from Theorem 11.5 and the following property of the rotation group $\mathcal{R}$, the proof of which we'll defer for just a moment.

Proposition 11.8. *$\mathcal{R}$ contains a free subgroup on two generators.*

What's being asserted here is:

> There exist two rotations $\rho, \sigma \in \mathcal{R}$ with the property that no nonempty reduced word in the "alphabet" $\mathscr{A} = \{\rho, \sigma, \rho^{-1}, \sigma^{-1}\}$ represents the identity transformation.

A "word" in the alphabet $\mathscr{A}$ is a string of symbols $x_1 x_2 \ldots x_n$ with each "letter" x_j an element of $\mathscr{A}$. Each such word "represents" the element of $\mathcal{R}$ obtained by viewing juxtaposition of letters as group multiplication (in this case, composition of mappings). As in the case of F_2, to say a word is "reduced" means that no letter stands next to its inverse.

Granting the above reformulation of the statement of Proposition 11.8, it's fortunate that only one reduced word can represent a given element of $\mathcal{R}$. Equivalently:

> Starting with a word composed of "letters" in the alphabet $\mathscr{A}$, the same reduced word results, no matter how the reduction is performed.

Proof. Suppose $v = x_1 x_2 \ldots x_m$ and $w = y_1 y_2 \ldots y_n$ are two different reduced words in the alphabet $\mathscr{A}$. We wish to prove that they multiply out to different group elements. We may without loss of generality assume that $x_m \neq y_n$ (else cancel these, and keep canceling rightmost letters until you first encounter ones that are distinct; this must happen eventually since $v \neq w$). Let g denote the element of $\mathcal{R}$ you get by interpreting v as a group product, and let $h \in \mathcal{R}$ correspond in this way to w. The word

$$z = vw^{-1} = x_1 x_2 \ldots x_m y_n^{-1} y_{n-1}^{-1} \ldots y_2^{-1} y_1^{-1}$$

corresponds to the group element gh^{-1}.

Claim. *z is a reduced word.*

For this, note that since v and w are reduced, the only cancellation possible in z is at the place where v and w^{-1} join up (w^{-1} is also reduced), i.e., at the pair $x_m y_n^{-1}$. But $x_m \neq y_n$, so no cancellation occurs there, either.

Since z is not the empty word, the property asserted above for the generators ρ and σ guarantees that gh^{-1} is not the identity element of $\mathcal{R}$, i.e., $g \neq h$, so different reduced words in the alphabet $\mathscr{A}$ must correspond to different group elements—as desired. □

Thus we can view the subgroup $\mathscr{F}$ of $\mathcal{R}$ generated by ρ and σ as giving an alternate construction of the free group F_2 on two generators; in particular, it's paradoxical!

Proof of Theorem 11.7. The subgroup $\mathscr{F}$ of $\mathscr{R}$ is countable, and each of its non-identity elements has exactly two fixed points (the points of intersection of the axis of that rotation with S^2). Thus the set C of these fixed points is countable, and by Exercise 11.5, $\mathscr{F}$ is a group of self-maps of $S^2 \backslash C$ that is fixed-point free on that set. Thus the Transference Theorem (Theorem 11.5) guarantees that $S^2 \backslash C$ is paradoxical for $\mathscr{F}$, and therefore also for $\mathscr{R}$.   □

Proof of Proposition 11.8. Choose $\rho \in \mathscr{R}$ to be rotation through $\theta = \sin^{-1}\left(\frac{4}{5}\right)$ radians about the z-axis and σ to be rotation through the same angle about the x-axis. We'll identify these maps with the matrices that represent them relative to the standard unit-vector basis of $\mathbb{R}^3$:

$$\rho = \begin{pmatrix} \frac{3}{5} & -\frac{4}{5} & 0 \\ \frac{4}{5} & \frac{3}{5} & 0 \\ 0 & 0 & 1 \end{pmatrix}, \qquad \sigma = \begin{pmatrix} 1 & 0 & 0 \\ 0 & \frac{3}{5} & -\frac{4}{5} \\ 0 & \frac{4}{5} & \frac{3}{5} \end{pmatrix}.$$

Since ρ and σ are orthogonal matrices their inverses are their transposes, so to say a reduced word of length n in these matrices and their inverses does not multiply out to the identity matrix is to say that the corresponding word in 5ρ, 5σ, and *their transposes* does not multiply out to 5^n times the identity matrix. For *this* it's enough to show that no such word multiplies out to a matrix all of whose entries are divisible by 5, i.e., that *over the field $\mathbb{Z}_5$ of integers modulo 5, no such word multiplies out to the zero-matrix!*

Over the field $\mathbb{Z}_5$ our matrices 5ρ, 5σ, and their transposes become

$$r = \begin{pmatrix} 3 & 1 & 0 \\ 4 & 3 & 0 \\ 0 & 0 & 0 \end{pmatrix}, \quad r' = \begin{pmatrix} 3 & 4 & 0 \\ 1 & 3 & 0 \\ 0 & 0 & 0 \end{pmatrix}, \quad s = \begin{pmatrix} 0 & 0 & 0 \\ 0 & 3 & 1 \\ 0 & 4 & 3 \end{pmatrix}, \quad s' = \begin{pmatrix} 0 & 0 & 0 \\ 0 & 3 & 4 \\ 0 & 1 & 3 \end{pmatrix}.$$

Let's call a word in the letters r, r', s, s' *admissible*[2] if r never stands next to r', and s never next to s'.

Our job now is to show that no admissible word in these new matrices multiplies out to the zero-matrix. We'll do this by identifying each matrix with the linear transformation it induces by left-multiplication on the (column) vector space $\mathbb{Z}_5^3$, and proving something more precise:

CLAIM. *The kernel of each admissible word in the letters r, r', s, s' is the kernel of its last letter.*

Proof of Claim. Each of the matrices r, r', s, s' has one dimensional range (i.e., column space) and two dimensional kernel. Upon calculating these ranges and kernels explicitly we find that the ranges of the "r-matrices" intersect the kernels of "s-

---

[2] We eschew the term "reduced" because, while in our original setup we had, e.g., $\rho\rho^{-1} = \rho^{-1}\rho = I$, now we have $rr' = r'r = 0$.

matrices" in $\{0\}$, and the same is true of the way the kernels of r-matrices intersect the ranges of s-matrices.

Now proceed by induction on word-length. The result is trivial for words of length one. Suppose $n \geq 1$ and that the kernel of each admissible word of length n in r, r', s, s' equals the kernel of that word's last letter. We wish to prove that the same is true of every admissible word of length $n+1$. Let w be such a word, so $w = va$ where v is an admissible word of length n and $a \in \{r, r', s, s'\}$. Then $x \in \ker w$ means that $vax = 0$, i.e., that $ax \in \ker v \cap \operatorname{ran} a$. Since w is an admissible word, the last letter of v, call it b, is not a', and by the induction hypothesis $\ker v = \ker b$. Thus $ax \in \ker b \cap \operatorname{ran} a = \{0\}$, so $x \in \ker a$. We've shown that $\ker w \subset \ker a$. The opposite inclusion is trivial, so $\ker w = \ker a$, as we wished to show.   □

Corollary 11.9. *The group $\mathscr{R}$ of rotations of $\mathbb{R}^3$ about the origin is paradoxical, hence not amenable.*

Proof. $\mathscr{R}$ inherits the paradoxicality of its subgroup $\mathscr{F}$ (Corollary 11.6, p. 134), hence it's not amenable (Exercise 11.1, p. 132).   □

11.3 Equidecomposability

According to Hausdorff's Paradox, if we remove a certain countable subset from S^2 then what remains is paradoxical with respect to $\mathscr{R}$, the group of rotations of $\mathbb{R}^3$ about the origin. In the next section we'll show, using an "absorption" technique similar the one used to prove Proposition 11.1, that S^2 itself is paradoxical with respect to $\mathscr{R}$. To do this efficiently it will help to have some new terminology.

For the rest of this section, G will denote a group of self-maps of a set X.

Definition 11.10 (Equidecomposability). *For subsets E and F of X: To say E is G-equidecomposable with F means that there exists a partition $\{E_i\}_1^n$ of E, a partition $\{F_i\}_i^n$ of F, and mappings $\{g_i\}_1^n \subset G$ such that $F_i = g_i E_i$ $(1 \leq i \leq n)$.*

Since the inverse of each map in G also belongs to G, it's clear that this notion of "equidecomposable" is symmetric: E is G-equidecomposable with F if and only if F is G-equidecomposable with E. In this case we'll just say "E and F are G-equidecomposable," and use the notation "$E \sim_G F$" to abbreviate the situation. Usually the group G is understood, in which case we'll just say "E and F are equidecomposable," and write $E \sim F$. If we wish to be more precise we'll say "E and F are equidecomposable using n pieces," and write $E \sim_n F$.

The notion of "equidecomposability" allows an efficient restatement of the definition of paradoxicality (Definition 11.2):

Proposition 11.11 (Definition of "Paradoxical" revisited). *A subset E of X is G-paradoxical if and only if there exists a partition of E into subsets A and B such that $A \sim E \sim B$.*

It's an easy exercise to show the relation "$\sim$" on $\mathscr{P}(X)$ is reflexive ($E \sim E$ for every $E \subset X$) and symmetric ($E \sim F \implies F \sim E$). In fact:

Theorem 11.12. *Equidecomposability is an equivalence relation.*

Proof. We need only prove transitivity. Suppose $E \sim F$ and $F \sim H$ for subsets E, F, H of X. Thus there exist partitions $\{E_i\}_1^n$ and $\{F_i\}_1^n$ of E and F, respectively, and transformations $\{g_i\}_1^n \subset G$ such that $g_i E_i = F_i$ for $1 \leq i \leq n$. There also exist partitions $\{F_j'\}_1^m$ and $\{H_j\}_1^m$ of F and H, respectively, and transformations $\{h_j\}_1^m \subset G$ such that $h_j F_j' = H_j$. Let $E_{i,j} = E_i \cap g_i^{-1}(F_i \cap F_j')$, and set $\gamma_{i,j} = h_j g_i$ on $E_{i,j}$. Thus each $\gamma_{i,j} \in G$, and one checks easily that (after removing empty sets, if necessary) $\{E_{i,j} : 1 \leq i \leq n, 1 \leq j \leq m\}$ and $\{\gamma_{i,j} E_{i,j} : 1 \leq i \leq n, 1 \leq j \leq m\}$ partition E and H, respectively. Thus $E \sim H$, as desired. $\qquad\square$

The notion of "same cardinality" is defined in terms of arbitrary bijections. In this vein, "equidecomposable" is a refinement of that concept, defined in terms of special bijections. More precisely:

Definition 11.13 (Puzzle Map). For subsets E and F of X, to say a bijection φ of E onto F is a *puzzle map* (more precisely: a "G-puzzle map") means that there is a partition $\{E_i\}_1^n$ of E and transformations $\{g_i\}_1^n \subset G$ such that $\varphi \equiv g_i$ on E_i.

The terminology suggests that we think of E as a jigsaw puzzle assembled from some finite collection of pieces, which the puzzle map φ reassembles into another jigsaw puzzle F. With this definition we have the following equivalent formulation of the notion of equidecomposability:

Proposition 11.14 (Equidecomposability via Puzzle Maps). *Subsets E and F of X are G-equidecomposable if and only if there is a G-puzzle map taking E onto F.*

The fact that G-equidecomposability is an equivalence relation can be explained in terms of puzzle maps: reflexivity means that the identity map is a puzzle map, symmetry means that the inverse of a puzzle map is a puzzle map, and the just-proved transitivity means that compositions of puzzle maps are puzzle maps.

The usefulness of equidecomposability stems from the next result, which asserts that paradoxicality is a property, not just of subsets of X, but actually of $\sim_G$ equivalence classes of subsets.

Corollary 11.15. *Suppose E and F are G-equidecomposable subsets of X. Then E is G-paradoxical if and only if F is G-paradoxical.*

Proof. By symmetry we need only prove one direction. Suppose E is G-paradoxical. Proposition 11.11 provides us with disjoint subsets A and B of E such that $A \sim E \sim B$. Since $E \sim F$ we're given a puzzle map φ mapping E onto F. Since φ is one-to-one, $A' = \varphi(A)$ and $B' = \varphi(B)$ are disjoint subsets of F, and since the restriction of a puzzle map is clearly a puzzle map we know that $A' \sim A$ and $B' \sim B$. Thus by transitivity: $A' \sim A \sim E \sim F$ and $B' \sim B \sim E \sim F$, hence $A' \sim F \sim B'$, so F is G-paradoxical by Proposition 11.11. $\qquad\square$

11.4 The Banach–Tarski Paradox for S^2 and $\mathbb{B}^3$

We know so far that if we remove a certain countable subset C from S^2, then the remainder $S^2 \backslash C$ is paradoxical for the group $\mathscr{R}$ of rotations of $\mathbb{R}^3$ about the origin. Aided by our work on equidecomposability, we can now give an efficient proof that S^2 itself is paradoxical. For this we'll build on the "absorption" idea that established the paradoxicality of the free group F_2 (Proposition 11.1). Here is the main tool:

Lemma 11.16 (The Absorption Lemma). *Suppose X is a set, E is a subset of X, and C is a countable subset of E. Suppose G is an uncountable group of self-maps of X that takes C into E and is fixed-point free on C. Then E and $E \backslash C$ are G-equidecomposable.*

Proof. The key here is to establish the following:

CLAIM. There exists $g \in G$ such that the family of sets $\{g^n(C) : n \in \mathbb{N} \cup \{0\}\}$ is pairwise disjoint.

Granting this: Let $C_\infty = \biguplus_{n=0}^{\infty} g^n(C)$. Then $C_\infty \subset E$ and, since the sets $g^n(C)$ are pairwise disjoint, $g(C_\infty) = C_\infty \backslash C$. Thus

$$E \backslash C = (E \backslash C_\infty) \uplus (C_\infty \backslash C) = (E \backslash C_\infty) \uplus g(C_\infty) \sim_G (E \backslash C_\infty) \uplus C_\infty = E$$

which establishes the theorem, showing in addition that only two pieces suffice.

Proof of Claim. It's enough to show that for some $g \in G$ we have $g^n(C) \cap C = \emptyset$ for each $n \in \mathbb{N}$. Indeed, once this has been established then given positive integers m and n with $n > m$ we'll have

$$g^n(C) \cap g^m(C) = g^m(g^{n-m}(C) \cap C) = g^m(\emptyset) = \emptyset.$$

Thus to finish the proof it's enough to show that the subset H of G, consisting of maps g for which $g^n(C) \cap C \neq \emptyset$ for some $n \in \mathbb{N}$, is at most countable; the existence of the desired $g \in G$ will then follow from the uncountability of G.

To this end, note that given c and c' in C there is at most one $h \in G$ with $h(c) = c'$. For if $h' \in G$ also takes c to c' then $h^{-1}h'$ fixes c, hence (because the action of G is fixed-point free on C) $h^{-1}h'$ is the identity map on X, i.e., $h = h'$. Now $h \in H$ if and only if there exist points c and c' in C and $n \in \mathbb{N}$ such that $h^n(c) = c'$. By the uniqueness just established, if $k \in G$ has the property that $k^m(c) = c'$ for some non-negative integer m, then $k = h^{n-m}$. Thus given the pair (c, c'), there's at worst a countable family of maps $h \in G$ for which some (integer) power of h takes c to c'. Since there are only countably many such pairs (c, c'), the set of all such maps h, i.e., the set H, is countable. $\qquad \square$

Theorem 11.17 (Banach–Tarski for S^2). *The unit sphere S^2 of $\mathbb{R}^3$ is $\mathscr{R}$-paradoxical.*

Proof. We know from the Hausdorff Paradox (Theorem 11.7) that S^2 contains a countable subset C such that $S^2 \backslash C$ is paradoxical. Choose a line L through the origin that does not intersect C, and let G denote the subgroup of $\mathscr{R}$ consisting of rotations with axis L. Thus G is an uncountable group that is fixed-point free on C, so the Absorption Lemma with $X = E = S^2$, tells us that S^2 is G-equidecomposable (hence also is $\mathscr{R}$-decomposable) with $S^2 \backslash C$. Corollary 11.15 now guarantees that S^2 inherits the $\mathscr{R}$-paradoxicality of $S^2 \backslash C$. $\square$

Theorem 11.17 gives an embryonic Banach–Tarski Paradox for the closed unit ball $\mathbb{B}^3$, i.e., the set of vectors in $\mathbb{R}^3$ of that lie at distance at most 1 from the origin.

Corollary 11.18. $\mathbb{B}^3 \backslash \{0\}$ is $\mathscr{R}$-paradoxical.

Proof. The $\mathscr{R}$-paradoxicality of S^2 means that it contains disjoint subsets A and B such that
$$A \sim S^2 \sim B \tag{11.2}$$
(Proposition 11.11). Let $A^* = \bigcup_{a \in A} \{ra : 0 < r \leq 1\}$, and similarly define B^*. Thus $\{A^*, B^*\}$ is a partition of $\mathbb{B}^3 \backslash \{0\}$, and $A^* \sim \mathbb{B}^3 \backslash \{0\} \sim B^*$ via the rotations responsible for (11.2). Thus $\mathbb{B}^3 \backslash \{0\}$ is $\mathscr{R}$-paradoxical. $\square$

Exercise 11.6. Show that both $\mathbb{R}^3 \backslash \mathbb{B}^3$ and $\mathbb{R}^3 \backslash \{0\}$ are $\mathscr{R}$-paradoxical.

The next exercise gives a nontrivial instance of the failure of the Transference Theorem (Theorem 11.5) if the action of the group G is not fixed-point free.

Exercise 11.7. Show that, with respect to the group $\mathscr{R}$ of rotations of $\mathbb{R}^3$ about the origin, $\mathbb{B}^3$ is *not* paradoxical.

Exercise 11.7 also shows that in order to establish the full Banach–Tarski Paradox for $\mathbb{B}^3$ we'll need to go beyond the group of rotations about the origin. Let $\mathscr{G}$ denote the group of rigid motions of $\mathbb{R}^3$ (i.e., the collection of isometric mappings taking $\mathbb{R}^3$ onto itself). In particular, every rotation, whether centered at the origin or not, belongs to $\mathscr{G}$.

Theorem 11.19 (Banach–Tarski for $\mathbb{B}^3$). *The three dimensional unit ball is a $\mathscr{G}$-paradoxical subset of $\mathbb{R}^3$.*

Proof. Let L be the line through the point $(0, 0, \frac{1}{2})$ parallel to the x-axis. Let $\mathscr{G}_L$ denote the subgroup of $\mathscr{G}$ consisting of rotations with axis L. Trivially $\mathscr{G}_L$ is fixed-point free on the singleton $\{0\}$, which it takes into $\mathbb{B}^3$. Upon setting $X = \mathbb{R}^3$, $E = \mathbb{B}^3$, and $C = \{0\}$ in the Absorption Lemma we see that $\mathbb{B}^3$ and $\mathbb{B}^3 \backslash \{0\}$ are $\mathscr{G}_L$-equidecomposable, hence $\mathscr{G}$-equidecomposable (using two pieces). Thus $\mathbb{B}^3$ inherits the $\mathscr{G}$-paradoxicality of $\mathbb{B}^3 \backslash \{0\}$. $\square$

Thus each closed ball can be thought of as a three dimensional jigsaw puzzle that can be reassembled, using only rotations (not all of them about the ball's center), into *two* closed balls of the same radius. This raises further questions: Is every ball $\mathscr{G}$-equidecomposable with every other ball? With a cube? We'll take up these matters in the next section.

Exercise 11.8. Continuing in the spirit of Exercise 11.2: in Definition 11.10 let's say that the sets E and F are *n-equidecomposable,* and write $E \sim_n F$.

(a) Show that $E \sim_m F$ and $F \sim_n H$ imply $E \sim_{mn} H$.

(b) Show that S^2 is 8-paradoxical with respect to the rotation group $\mathscr{R}$.

(c) Show that $\mathbb{B}^3$ is 16-paradoxical with respect to the isometry group $\mathscr{G}$.

11.5 Banach–Tarski beyond $\mathbb{B}^3$

Galileo in 1638 discussed the paradox one encounters in comparing the sizes of infinite sets. Using the notation "$A \sim B$" for "there exists a bijection of set A onto set B" (i.e., "A and B have the same cardinality") Galileo's Paradox can be expressed as follows:

If $\mathbb{N}$ is the set of natural numbers, S the subset of squares, and T the subset of nonsquares, then, even though $\mathbb{N}$ is the disjoint union of S and T, it's nonetheless true that $S \sim \mathbb{N} \sim T$.

Proposition 11.11 phrases the notion of paradoxicality in similar terms, but now using the more sophisticated equivalence relation of "equidecomposability." Like the notion of "same cardinality," equidecomposability can be defined in terms of bijections, but now the bijections are "piecewise congruences," i.e., *puzzle maps* (Proposition 11.14).

The deepest elementary result about "same cardinality" is the Schröder–Bernstein Theorem: if set A has the same cardinality as a subset of set B, and B has the same cardinality as a subset of A, then A and B have the same cardinality. The same is true for equidecomposability; the two results even have a common proof! In this section we'll give this proof and examine its astonishing consequences for the notion of paradoxicality.

We'll assume as usual that G is a group of self-maps of a set X, and we'll continue to write $A \sim_G B$ for "A and B are G-equidecomposable."

Notation 11.20. By "$A \preceq_G B$" we mean "A is G-equidecomposable with a subset of B," i.e., "There is a puzzle map taking A onto a subset of B."

Thus the relation $\preceq_G$ is *reflexive* since the identity map is a puzzle map, and *transitive* since the composition of puzzle maps is a puzzle map. To proceed further we'll need a simple observation about the ordering $\preceq_G$.

Lemma 11.21. *Suppose $\{A_j\}_1^n$ and $\{B_j\}_1^n$ are families of subsets of X, each of which is pairwise disjoint.*

(a) *If $A_j \preceq_G B_j$ for each index j, then $\biguplus_{j=1}^n A_j \preceq_G \biguplus_{j=1}^n A_j$.*

(b) *If $A_j \sim_G B_j$ for each index j, then $\biguplus_{j=1}^n A_j \sim_G \biguplus_{j=1}^n A_j$.*

Proof. (a) Our hypothesis is that for each j there is a puzzle map φ_j taking A_j into B_j. Then it's easy to check that the map φ defined by setting $\varphi = \varphi_j$ on A_j is a puzzle map taking the union of the A_j's onto the union of the B_j's.

(b) Same as (a), except now the puzzle map φ_j takes A_j *onto* B_j $(1 \leq j \leq n)$, and therefore φ takes $\uplus_1^n A_j$ *onto* $\uplus_1^n B_j$. $\square$

The key to the rest of this section is the fact that $\preceq_G$, in addition to being reflexive and transitive, is also *antisymmetric*, and so induces a partial order on $\mathscr{P}(X)$. This is the content of:

Theorem 11.22 (The Banach–Schröder–Bernstein Theorem). *If A and B are subsets of X with $A \preceq_G B$ and $B \preceq_G A$, then $A \sim_G B$.*

Proof. The hypotheses assert that there are puzzle maps f and g with f taking A onto a subset B_1 of B and g taking B onto a subset A_1 of A. By the Banach Mapping Theorem (Theorem 1.1, p. 8) there is a subset C of A such that g takes $B \backslash f(C)$ *onto* $A \backslash C$. Since g is a puzzle map, and since the restriction of a puzzle map is again a puzzle map, this equation asserts that $B \backslash f(C) \sim A \backslash C$, where here—and in the arguments to follow—we allow ourselves to omit the subscript $\mathscr{G}$. Since f is a puzzle map we know that $f(C) \sim C$. Thus Lemma 11.21 insures that

$$B = (B \backslash f(C)) \cup f(C) \sim (A \backslash C) \cup C = A$$

as desired. $\square$

Previously we noted that for subsets A and B of X:

$$A \subset B \implies A \preceq_G B.$$

Recall the notation $\mathscr{G}$ for the group of all isometric self-maps of $\mathbb{R}^3$.

Corollary 11.23. *Suppose $\{B_j\}_1^n$ is a pairwise disjoint family of subsets of $\mathbb{R}^3$, each of which is $\mathscr{G}$-equidecomposable with $\mathbb{B}^3$. Then $\uplus_{j=1}^n B_j \sim_{\mathscr{G}} \mathbb{B}^3$.*

Proof. We proceed by induction on n; if $n = 1$ there is nothing to prove, so suppose $n > 1$ and that the result is true for $n - 1$. Let $C_1 = \uplus_{j=1}^{n-1} B_j$ and $C_2 = C_1 \uplus B_n$; our goal is to show that $C_2 \sim \mathbb{B}^3$. Now both C_1 (induction hypothesis) and B_n are $\sim \mathbb{B}^3$ and by the Banach–Tarski Theorem there exists a partition $\{E_1, E_2\}$ of $\mathbb{B}^3$ such that E_1 and E_2 are each $\sim \mathbb{B}^3$. Thus $E_1 \sim C_1$ and $E_2 \sim B_n$, so by Lemma 11.21, $\mathbb{B}^3 = E_1 \uplus E_2 \sim C_1 \uplus B_n = C_2$. $\square$

Corollary 11.24. *Every closed ball in $\mathbb{R}^3$ is $\mathscr{G}$-equidecomposable with every other closed ball.*

Proof. Fix a closed ball B in $\mathbb{R}^3$. It's enough to prove that B is equidecomposable with the closed unit ball $\mathbb{B}^3$.

Suppose first that the radius of B is > 1. Cover B by balls $\{B_j\}_1^n$ of radius equal to one, and "disjointify" this collection of B_j's by setting

$$B_j' = B_j \backslash \cup_{k=j+1}^n B_k.$$

Then $B'_j \subset B_j$ for each index j and the new collection $\{B'_j\}_1^n$ has the same union as the original one; in particular it still covers B. Now let $\{C_j\}_1^n$ be a pairwise disjoint collection of closed balls of radius 1 in $\mathbb{R}^3$. Then

$$\mathbb{B}^3 \preceq B \preceq \biguplus_{j=1}^n B'_j \preceq \biguplus_{j=1}^n C_j \sim \mathbb{B}^3$$

where the first "inequality" comes from the containment of $\mathbb{B}^3$ in a translate of B, the second one from the containment of B in the union of the B'_j s and the third one from Lemma 11.21 above along with the containment of each B'_j in a translate of the corresponding C_j. Corollary 11.23 provides the final "equality." Thus $\mathbb{B}^3 \preceq B \preceq \mathbb{B}^3$, so $B \sim \mathbb{B}^3$ by the Banach–Schröder–Bernstein Theorem.

If the radius of B is < 1, repeat the above argument with the roles of B and $\mathbb{B}^3$ reversed. If the radius of B is equal to 1 then B, being a translate of $\mathbb{B}^3$, is trivially $\mathcal{G}$-equidecomposable with that set. $\qquad\square$

Corollary 11.25 (The "Ultimate" Banach–Tarski Theorem). *Every two bounded subsets of $\mathbb{R}^3$ with nonempty interior are $\mathcal{G}$-equidecomposable.*

Proof. Let E be a bounded subset of $\mathbb{R}^3$ with nonempty interior. It's enough to show that $E \sim \mathbb{B}^3$. Since E contains a closed ball B we know from Corollary 11.24 that $\mathbb{B}^3 \sim B \preceq E$. Since E is bounded it is contained in a closed ball B', so again by Corollary 11.24: $E \preceq B' \sim \mathbb{B}^3$, hence $\mathbb{B}^3 \preceq E \preceq \mathbb{B}^3$. Thus $E \sim \mathbb{B}^3$ by Banach–Schröder–Bernstein. $\qquad\square$

Exercise 11.9 (Paradoxicality revisited). In many expositions of the Banach–Tarski paradox the definition of "paradoxical" is taken to be somewhat less restrictive than the one we've used here (Definition 11.2). Specifically: the "replicator family" $\{E_n\}_1^n$ of that definition is often required only to be pairwise disjoint (not necessarily with union equal to E), while the "replicant families," although still required to exhaust all of E, no longer need to be pairwise disjoint. Show that in this revised definition:

(a)   The replicant families can, without loss of generality, be assumed to be pairwise disjoint. Thus we can rephrase the new definition as follows: *There exist disjoint subsets A_0 and B contained in E with $A_0 \sim E \sim B$.*

(b)   Let $A = A_0 \uplus E \backslash (A_0 \cap B)$, so that $E = A \uplus B$. Show that $A \sim E$. Thus the new "weakened" definition of paradoxicality is equivalent to the original one.

Notes

A free group of rotations. The idea to consider the two matrices used in the proof of Proposition 11.8, and to transfer the argument to the field $\mathbb{Z}_5$, comes from Terry Tao's intriguing preprint [115].

The Hausdorff Paradox. The original version of Hausdorff's Paradox occurs in [48]; it asserts that there exists a countable subset C of S^2 such that $S^2 \backslash C$ can be partitioned into three subsets A, B, and C such that each is congruent via rotations to the others, *and also to $B \cup C$*. Hausdorff's motivation here was to show that $\mathscr{P}(S^2)$ does not support a rotation-invariant finitely additive probability measure.

References for the Banach–Tarski Paradox. The results of Sects. 11.3–11.5 all come from Banach and Tarski's famous paper [10]. See Chap. 3 of Stan Wagon's book [121] (the gold standard for exposition on the Banach–Tarski Paradox and the research it has inspired up through 1992) for more on the material we've covered here. See also [104, Chap. 1] for another exposition of the Banach–Tarski Paradox and for more recent developments, with the emphasis on amenability. A more popularized exposition of the Banach–Tarski paradox is Leonard Wapner's delightful book [122], which provides much interesting biographical information about the personalities involved, as well as commentary on the foundational issues raised by this amazing theorem.

"n-Paradoxicality." Regarding Exercises 11.2 and 11.8: Raphael Robinson proved in the 1940s that S^2 is 4-paradoxical with respect to the rotation group $\mathscr{R}$ and that $\mathbb{B}^3$ is 5-paradoxical (but not 4-paradoxical) with respect to the full isometry group $\mathscr{G}$. Wagon discusses these matters, with appropriate references, in Chap. 4 of [121].

Amenability and paradoxicality. We've seen that paradoxical groups are not amenable (Exercise 11.1). In the late 1920s Tarski proved that the converse is true: *If a group is not amenable, then it is paradoxical.* See [121, Chap. 9, pp. 125–129] for an exposition of this remarkable theorem.

Galileo's Paradox. In his treatise [41] (pp. 31–33) Galileo observes that the set of squares in $\mathbb{N}$ is in one-to-one correspondence with $\mathbb{N}$ itself, and so has the same size as $\mathbb{N}$. He concludes that size comparisons between infinite sets are impossible.

Chapter 12
Fixed Points for Non-commuting Map Families

MARKOV–KAKUTANI FOR SOLVABLE FAMILIES

Overview. Here we'll generalize the Markov–Kakutani Theorem (Theorem 9.6, p. 107) to collections of affine, continuous maps that obey a generalized notion of commutativity inspired by the group-theoretic concept of solvability. This will enable us to show, for example, that the unit disc is not paradoxical even with respect to its *full* isometry group, and that solvable groups are amenable, hence not paradoxical. We'll prove that compact solvable groups possess Haar measure, and will show how to extend this result to solvable groups that are just *locally* compact.

12.1 The "Solvable" Markov–Kakutani Theorem

We know from the Banach–Tarski Paradox (Theorem 11.19) that $\mathbb{B}^3$ is paradoxical with respect to the full isometry group of $\mathbb{R}^3$. Thanks to the Markov–Kakutani Fixed-Point Theorem and the commutativity of the group of origin-centered rotations of $\mathbb{R}^2$, we also know (Corollary 10.5, p. 123) that there is defined, for all subsets of the unit disc $\mathbb{B}^2$, a finitely additive probability measure that is rotation-invariant. Consequently (Exercise 11.1) $\mathbb{B}^2$ is not paradoxical with respect to the group of rotations of $\mathbb{R}^2$ about the origin. This raises the question:

Is $\mathbb{B}^2$ paradoxical with respect to its full *group of isometries?*

The isometry group of $\mathbb{B}^2$ allows, in addition to rotations about the origin, *reflections* in a line through the origin; this creates non-commutativity. Indeed, we know from linear algebra that the rotations of $\mathbb{R}^2$ about the origin are the linear transformations represented (with respect to the standard unit-vector basis of $\mathbb{R}^2$) by matrices of the form $\left[\begin{smallmatrix} \cos\theta & -\sin\theta \\ \sin\theta & \cos\theta \end{smallmatrix}\right]$, where $\theta \in [0, 2\pi)$ is the angle of rotation. These rotation matrices form the subgroup SO(2) of O(2), the group of all 2×2 matrices whose columns form an orthonormal set in $\mathbb{R}^2$. Each matrix in O(2) has determinant ± 1 (a consequence of column-orthonormality, which can be rephrased: "The transpose of each matrix in O(2) is its inverse"), those with determinant -1 being the *reflections*

© Springer International Publishing Switzerland 2016   145
J.H. Shapiro, *A Fixed-Point Farrago*, Universitext,
DOI 10.1007/978-3-319-27978-7_12

about lines through the origin, and those with determinant $+1$ constituting the rotation group $SO(2)$. The isometries of $\mathbb{R}^2$ that fix the origin are precisely the linear transformations represented by matrices in $O(2)$. More generally the same is true for $\mathbb{R}^N$, with $O(N)$ in place of $O(2)$ (see Appendix D for the full story). Now the matrix group $O(2)$ (hence its *alter ego*, the isometry group of the unit disc) is not commutative, as witnessed by the pair of matrices:

$$\begin{bmatrix} \frac{1}{\sqrt{2}} & -\frac{1}{\sqrt{2}} \\ \frac{1}{\sqrt{2}} & \frac{1}{\sqrt{2}} \end{bmatrix} \quad \text{and} \quad \begin{bmatrix} 1 & 0 \\ 0 & -1 \end{bmatrix},$$

the first of which induces rotation through an angle of 45 degrees about the origin, while the second induces reflection about the horizontal axis.

In this chapter we'll generalize the Markov–Kakutani Theorem in a way that applies to non-commutative groups like $O(2)$; in so doing we'll be able to extend the disc's non-paradoxicality from rotations to *all* its isometries.

Theorem 12.1 (The "Solvable" Markov–Kakutani Theorem). *Suppose K is a non-void compact, convex subset of a Hausdorff topological vector space. Then every solvable family of continuous, affine self-maps of K has a common fixed point.*

We'll devote the next section to understanding the meaning of "solvable," after which we'll prove Theorem 12.1 and show how to apply it.

12.2 Solvable Families of Maps

Our notion of solvability is inspired by group theory (see Appendix E).

Definition 12.2. Suppose $\mathscr{A}$ is a family of self-maps of some set.

(a)   *Solvable family of maps.* This is what we'll call $\mathscr{A}$ whenever there is a finite chain of subfamilies

$$\{\text{Identity map}\} \;=\; \mathscr{A}_0 \subset \mathscr{A}_1 \subset \mathscr{A}_2 \subset \cdots \subset \mathscr{A}_n = \mathscr{A} \qquad (12.1)$$

such that for each $1 \leq k \leq n$ and each pair A, B of maps in $\mathscr{A}_k$ there exists a "commutator" $C \in \mathscr{A}_{k-1}$ such that $AB = BAC$.

(b)   *Solvability degree.* More precisely, we may call $\mathscr{A}$ as above "n-solvable."

(c)   *Solvable group.* This is what we'll call the family $\mathscr{A}$ whenever it satisfies condition (a) above, and each of the subfamilies $\mathscr{A}_k$ in (12.1) is a *group* under composition. For more precision we may use the term "n-solvable group."

Remarks 12.3. Suppose $\mathscr{A}$ denotes a family of self-maps that is solvable in the sense of Definition 12.2 (a).

(a) *Solvability and commutativity.* $\mathscr{A}_1$ is commutative, so $\mathscr{A}$ is "1-solvable" if and only if it is commutative. "2-solvable" is the next-best thing, ...

(b) *Semigroups and groups of self-maps.* For a family of self-maps of a set, the collection of common fixed points is not changed if one replaces the original family of self-maps by the "unital semigroup" it generates, i.e., the set of all possible finite compositions of the original maps, along with the identity map. If each map of the original family is a bijection, we can even add all the inverses to the original family without changing the common fixed-point set, in which case the new "inverse-enhanced" family generates a *group* under composition having the same common fixed-point set as the original family. Here we'll only consider self-map families that are groups.

(c) *Solvable groups.* Suppose G is a group with identity element e. We can consider G to be a group of self-maps, acting itself by (say) left multiplication. Each pair of elements $a, b \in G$ has a unique commutator $[a;b] := (ba)^{-1}ab = a^{-1}b^{-1}ab$. Thus, according to Definition 12.2(c) above: G is a *solvable group* if and only if there is a chain of *subgroups*

$$\{e\} = G_0 \subset G_1 \subset G_2 \ldots G_n = G \tag{12.2}$$

such that for each index k between 1 and n the subgroup G_{k-1} contains all the commutators of G_k.

The usual definition of "solvable" for groups stipulates that for $1 \leq k \leq n$ the subgroup G_{k-1} must to be a *normal* subgroup of G_k, and that furthermore each quotient group G_k/G_{k-1} must be abelian. These requirements of normality plus commutativity turn out to be equivalent to the single commutator-containment condition of the last paragraph; see Appendix E for the details.

Example 12.4. The matrix group $O(2)$ *is solvable.* We've noted that the family of isometric self-maps of $\mathbb{B}^2$ can be identified with $O(2)$ acting by left-multiplication of column vectors. Consider the chain of subgroups

$$\{I\} \subset SO(2) \subset O(2), \tag{12.3}$$

noting that $SO(2)$, the group of 2×2 rotation matrices, is commutative. The multiplicative property of determinants now takes over; each matrix in $O(2)$ has determinant either $+1$ or -1, and so has the same determinant as its inverse. Thus given matrices A and B in $O(2)$ the commutator $[A;B]$ belongs to $O(2)$ and has determinant $+1$; it therefore belongs to $SO(2)$.

Conclusion: $O(2)$ is a 2-solvable group in the sense of Definition 12.2.

Example 12.5. The affine group of $\mathbb{R}$ *is solvable.* Let $A(\mathbb{R})$ denote the collection of affine transformations of the real line, i.e., the transformations $\gamma_{r,t} : x \to rx + t$ ($x \in \mathbb{R}$) for $t \in \mathbb{R}$ and $r \in \mathbb{R}\backslash\{0\}$. Then, with composition as the binary operation in $A(\mathbb{R})$:

(a) $\gamma_{r,t} \circ \gamma_{\rho,\tau} = \gamma_{r\rho,r\tau+t}$, so $A(\mathbb{R})$ is a group, with $\gamma_{r,t}^{-1} = \gamma_{1/r,-t/r}$.

(b) $A(\mathbb{R})$ is generated by two commutative subgroups: the dilation group consisting of maps $\gamma_{r,0}$ where $r \neq 0$, and the translation group $T(\mathbb{R})$ consisting of maps $\gamma_{1,t}$ for $t \in \mathbb{R}$.

(c)   The commutator $[\gamma_{r,t};\gamma_{\rho,\tau}]=\gamma_{1,s}\in T(\mathbb{R})$, where $s=\frac{(1-\rho)t-(1-r)\tau}{r\rho}$.

Thus we have the chain of groups: $\{\text{identity map}\}\subset T(\mathbb{R})\subset A(\mathbb{R})$, where $T(\mathbb{R})$ is commutative and contains all the commutators of $A(\mathbb{R})$.

Conclusion: $A(\mathbb{R})$ is a 2-solvable group.

Exercise 12.1. Show that the map $\gamma_{r,t}\to\left[\begin{smallmatrix}r & t\\0 & 1\end{smallmatrix}\right]$ is a homomorphism taking $A(\mathbb{R})$ onto a group of invertible 2×2 real matrices, and that all the calculations in the example above can be done "matricially."

The exercises below give two more examples of solvable matrix groups, the second of which is 3-solvable, but not 2-solvable.

Exercise 12.2 (The Heisenberg group is solvable). The *Heisenberg group* is the collection $\mathscr{H}=\mathscr{H}_3(\mathbb{R})$ of 3×3 real matrices that are upper triangular and whose main diagonal consists entirely of 1's.

(a)   Show that $\mathscr{H}$ is a group under matrix multiplication.

(b)   Let $\mathscr{K}$ denote the subset of $\mathscr{H}$ consisting of matrices of the form $\left(\begin{smallmatrix}1 & 0 & a\\0 & 1 & 0\\0 & 0 & 1\end{smallmatrix}\right)$. Show that $\mathscr{K}$ is a commutative subgroup of $\mathscr{H}$.

(c)   Show that if $A,B\in\mathscr{H}$, then the commutator $A^{-1}B^{-1}AB$ belongs to $\mathscr{K}$. Conclude that $\mathscr{H}$ is 2-solvable.

Exercise 12.3 (The Upper-Triangular group is 3-solvable, but not 2-solvable). Let $\mathscr{U}$ denote the collection of 3×3 matrices that are *upper triangular,* i.e., have all entries zero below the main diagonal.

(a)   Show that $\mathscr{U}$ is a group under matrix multiplication.

(b)   Show that the Heisenberg group contains every $\mathscr{U}$-commutator. Conclude that $\mathscr{U}$ is 3-solvable.

(c)   Show that the $\mathscr{U}$ is not 2-solvable.

(d)   *Suggestion:* By considering, e.g., matrices of the form $A=\left(\begin{smallmatrix}1 & 1 & 0\\0 & 1 & 1\\0 & 0 & a\end{smallmatrix}\right)$ and $B=\left(\begin{smallmatrix}b & 0 & c\\0 & 1 & 0\\0 & 0 & d\end{smallmatrix}\right)$, show that the collection of commutators of $\mathscr{U}$ exhausts the *entire* Heisenberg group. Argue that if $\mathscr{U}$ were 2-solvable, then the Heisenberg group would have to be commutative, which it is not.

The next exercise concerns a famous class of finite groups that are *not* solvable.

Exercise 12.4. S_n *is not solvable for* $n\geq 5$. Here S_n denotes the set of permutations (1-to-1 onto maps) of a set of n elements, which we might as well take to be $[1,n]:=\{1,2,\ldots n\}$. With composition as its binary operation, S_n is a group (the symbol "S" stands for "symmetric"). We assume here that $n\geq 5$.

Of particular interest to us are the *3-cycles* in S_n, i.e., the maps that permute a triple $\{a,b,c\}$ of distinct elements of $[1,n]$ *cyclically:* $a\to b\to c\to a$ and leave everything else alone. Notation for such a 3-cycle: (a,b,c). In the exercises below we assume $n\geq 5$.

(a)   Show that the 3-cycle $(1,4,3)$ is the commutator $[\sigma,\tau]$ where $\sigma=(1,2,3)$ and $\tau=(3,4,5)$.

(b)   By making appropriate substitutions in part (a) show that *every* 3-cycle in S_n is a commutator of other 3-cycles.

(c)   Use part (b) to show that S_n is not solvable.

12.3 Proof of the solvable Markov–Kakutani Theorem

The argument proceeds by induction on the "solvability index" n in (12.1). Since $\mathscr{A}_1$ is commutative, the case $n = 1$ is just the original Markov–Kakutani Theorem (Theorem 9.6, p. 107).

For the induction step suppose $n \geq 2$ and the result is true for all $(n-1)$-solvable families, of which $\mathscr{A}_{n-1}$ in (12.1) is one. The set K_{n-1} of common fixed points for $\mathscr{A}_{n-1}$ is nonempty (induction hypothesis), compact (continuity of the maps in $\mathscr{A}_{n-1}$), and convex (affine-ness of the maps in $\mathscr{A}_{n-1}$).

Claim: Each map $A \in \mathscr{A} = \mathscr{A}_n$ takes K_{n-1} into itself.

Proof of Claim. Given $A \in \mathscr{A}$ and $p \in K_{n-1}$ we're claiming that $A(p) \in K_{n-1}$, i.e., that $BA(p) = A(p)$ for every $B \in \mathscr{A}_{n-1}$. Given $A \in \mathscr{A}$ and $B \in \mathscr{A}_{n-1}$ there exists $C \in \mathscr{A}_{n-1}$ such that $BA = ABC$. Thus for $p \in K_{n-1}$ we have (since both B and C belong to $\mathscr{A}_{n-1}$): $BA(p) = ABC(p) = AB(p) = A(p)$, as desired.

To finish the proof of Theorem 12.1 we're going to show that $\tilde{\mathscr{A}}$, the collection of restrictions to K_{n-1} of maps in $\mathscr{A}$, is commutative. This, along with the just-proved *Claim*, will establish $\tilde{\mathscr{A}}$ as a commutative family of continuous, affine self-maps of K_{n-1}. The original Markov–Kakutani Theorem will then provide for $\tilde{\mathscr{A}}$ a common fixed point $p \in K_{n-1}$, *a fortiori* a fixed point for every map in $\mathscr{A}$.

It remains to establish the desired commutativity for $\tilde{\mathscr{A}}$. For this, suppose A and B belong to $\mathscr{A}$ and choose $C \in \mathscr{A}_{n-1}$ so that $AB = BAC$. Then for $p \in K_{n-1}$ (hence a fixed point for C): $A(B(p)) = B(A(C(p))) = B(A(p))$, i.e., $AB = BA$ on K_{n-1}.   □

12.4 Applying the solvable M–K Theorem

Recall the cast of characters that emerged in Chaps. 9 and 10 when we applied the original the Markov–Kakutani Theorem.

(a)   There was a set S and a commutative family Φ of self-maps of S.

(b)   Each $\varphi \in \Phi$ gave rise to the (linear) composition operator $C_\varphi : f \to f \circ \varphi$ acting on $B(S)$ (the vector space of bounded, real-valued functions on S). We denoted the collection of all such composition operators by C_Φ.

These actors will return in this chapter, except that now we'll allow Φ to be "solvable" in the sense of Definition 12.2. The Markov–Kakutani triple $(X, K, \mathscr{A})$ of Sects. 9.5 and 10.1 will return unchanged:

(c) $X = B(S)^\sharp$, the algebraic dual of $B(S)$, taken in its *weak-star topology*.

(d) $K = \mathcal{M}(S)$, the set of "means" on $B(S)$, i.e., those positive linear functionals
       on $B(S)$ that take value 1 on the function $\equiv 1$ on S.

(e) $\mathcal{A} = C_\Phi^\sharp$, the collection of adjoints of composition operators belonging to the
       family C_Φ.

To apply our enhanced Markov–Kakutani Theorem we need only to show that the
solvability assumed for the original family Φ of self-maps of S is inherited by the
family $C_\Phi^\sharp$ of affine self-maps of $\mathcal{M}(S)$. For this one need only check that the map
$\varphi \to C_\varphi$ reverses composition $(C_{\varphi \circ \psi} = C_\psi C_\varphi)$, and that the same is true of the map
$T \to T^\sharp$ that associates to each linear transformation on a vector space its adjoint.
Thus the map $\varphi \to C_\varphi^\sharp$ preserves the order of composition; in particular, if $\varphi, \psi, \gamma \in$
Φ and γ is a commutator of the pair (φ, ψ) in the sense that $\varphi \circ \psi = \psi \circ \varphi \circ \gamma$,
then $C_\gamma^\sharp$ is a commutator of the pair $(C_\varphi^\sharp, C_\psi^\sharp)$. Consequently, if the original family of
maps Φ is solvable, then so is $C_\Phi^\sharp$. Theorem 12.1 can therefore be applied to yield

Theorem 12.6 (Invariant means for solvable families of maps). *Suppose Φ is a*
solvable family of self-maps of a set S. Then:

(a)   *There is a mean Λ on $B(S)$ that is invariant for $C_\Phi^\sharp$, i.e., $\Lambda \circ C_\varphi = \Lambda$ for every*
       $\varphi \in \Phi$ *(cf. Theorem 9.19).*

(b)   *There is a finitely additive Φ-invariant probability measure on $\mathscr{P}(S)$ (cf. The-*
       orem 10.3).

(c)   *S is not Φ-paradoxical (see Exercise 11.1).*

Since $\Phi = O(2)$ is a solvable family of self-maps of $\mathbb{B}^2$ and S^1, we see in particular:

Corollary 12.7. *$\mathbb{B}^2$ and S^1 are not $O(2)$-paradoxical.*

Corollary 12.8. *Solvable groups are amenable, hence not paradoxical.*

In the other direction we have

Corollary 12.9. *The following groups are not solvable:*

(a)   *The free group F_2 on two letters.*

(b)   *The compact group $SO(3)$ of 3×3 orthogonal matrices with determinant one.*

Proof. Both groups are not amenable (Theorem 10.12 for F_2 and Corollary 11.9 for
$SO(3)$ in its guise as the rotation group $\mathscr{R}$), hence not solvable.   □

If our basic set S is a compact topological space, then we have the following
extension of Corollaries 9.20 and 9.21:

Corollary 12.10. *If Φ is a solvable family of continuous affine self-maps of a com-*
pact topological space S, then there exists a regular Borel probability measure μ for
S such that

$$\int f \circ \varphi \, d\mu = \int f \, d\mu$$

for every $f \in C(S)$ and $\varphi \in \Phi$.

Corollary 12.11. *Every solvable compact topological group has a Haar measure.*

12.5 The (solvable) Invariant Hahn–Banach Theorem

Given a solvable family Φ of *continuous* self-maps of a *compact Hausdorff space S*, our solvably enhanced version of the Markov–Kakutani Theorem produces—just as did the original version in Sect. 10.2—two important Φ-invariant set functions for S: A regular probability measure μ on the Borel sets of S (Corollary 12.10), and a finitely additive probability measure ν defined for all subsets of S (Theorem 12.6). This brings up the same question we faced in Sect. 10.2: "Can ν be realized as an extension of μ?" Once again the answer is "yes," with the heavy lifting done by a "solvable" extension of the Invariant Hahn–Banach Theorem (Theorem 10.6, p. 124).

Theorem 12.12 (The "solvable" Invariant Hahn–Banach Theorem). *Suppose V is a vector space and $\mathscr{G}$ is a solvable family of linear transformations $V \to V$. Suppose W is a linear subspace of V that is taken into itself by every transformation in $\mathscr{G}$, and that p is a gauge function on V that is "$\mathscr{G}$-subinvariant" in the sense that*

$$p(\gamma(v)) \leq p(v) \text{ for every } v \in V \text{ and } \gamma \in \mathscr{G}.$$

Suppose Λ is a $\mathscr{G}$-invariant functional on W that is dominated by p, i.e.,

$$\Lambda \circ \gamma = \Lambda \text{ for all } \gamma \in \mathscr{G} \quad \text{and} \quad \lambda(v) \leq p(v) \text{ for all } v \in W.$$

Then Λ has a $\mathscr{G}$-invariant linear extension to V that is dominated on V by p.

Corollary 12.13. *If S is a compact Hausdorff space upon which acts a solvable family Φ of continuous self-maps, then each regular Φ-invariant probability measure on the Borel sets of S extends to a finitely additive probability measure defined for all subsets of S.*

The proofs of these two results are identical to those of their commutative analogues (Theorem 10.6 and Corollary 10.7, pp. 124–125), except that the solvable Markov–Kakutani Theorem replaces the original one.

We saw in Corollary 10.5 that for the closed unit disc $\mathbb{B}^2$ there is a finitely additive, rotation-invariant probability measure on $\mathscr{P}(\mathbb{B}^2)$. Thanks to Example 12.4 and Theorem 12.1 we now know there exists a finitely additive probability measure on $\mathscr{P}(\mathbb{B}^2)$ invariant for the *full* isometry group $O(2)$ of $\mathbb{B}^2$. Corollary 12.13 shows that this isometry-invariant finitely additive probability measure can be chosen to extend normalized Lebesgue area measure; similar results hold for the unit circle.

Invariant extension of Lebesgue measure on $\mathbb{R}$. Lebesgue measure m on the Borel subsets of the real line is invariant under translations, and "scales properly" under dilations. More precisely: for each pair (r,t) of real numbers, and each Borel subset E of $\mathbb{R}$, we have $m(rE+t) = |r|m(E)$. Thanks to the solvability of the affine group $A(\mathbb{R})$ of the real line (Exercise 12.5), our "solvable" Invariant Hahn–Banach Theorem provides an extension of Lebesgue measure to a finitely additive measure on $\mathscr{P}(\mathbb{R})$ that preserves the translation-invariance and scaling properties of the original. More precisely:

Theorem 12.14. *There is an extension of Lebesgue measure to a finitely additive measure μ on all subsets of $\mathbb{R}$ such that for each $E \subset \mathbb{R}$:*

$$\mu(rE + t) = |r|\mu(E) \qquad (0 \neq r \in \mathbb{R}, \, t \in \mathbb{R}) \tag{12.4}$$

and

$$m_*(E) \leq \mu(E) \leq m^*(E), \tag{12.5}$$

where $m_*(E)$ and $m^*(E)$ denote, respectively, the inner and outer measures of E.

Proof. Let V be the collection of real-valued functions f on $\mathbb{R}$ for which the upper integral

$$\int^* |f| = \inf \left\{ \int s : s \in \mathscr{S}, |f| \leq s \right\}$$

of $|f|$ over $\mathbb{R}$ is finite; here $\mathscr{S}$ denotes the collection of Borel-measurable, integrable functions on $\mathbb{R}$ that are *simple*, i.e., take only finitely many values, and the integrals are taken with respect to Lebesgue measure on the line. For $f \in V$ let $p(f) = \int^* |f|$.

> Exercise 12.5. Prove that p is a gauge function on V, as defined in the statement of the Hahn–Banach Theorem on p. 124.

For $\gamma = \gamma_{r,t} \in A(\mathbb{R})$ (notation as in Example 12.5), define the linear transformation L_γ on V as the "weighted" composition operator:

$$(L_\gamma f)(x) = rf(\gamma(x)) \qquad (f \in V, \, x \in \mathbb{R}),$$

and let $\mathscr{A}$ denote the collection of all such transformations. With composition as its binary operation, $\mathscr{A}$ is a group that inherits the 2-solvability of $A(\mathbb{R})$, and so satisfies the hypotheses of the "solvable" Invariant Hahn–Banach Theorem.

Thanks to the change-of-variable formula for Lebesgue integrals, the functional p is invariant for each $L_\gamma \in \mathscr{A}$:

$$p(L_\gamma f) = p(f) \qquad (f \in V, \, \gamma \in A(\mathbb{R})).$$

Let W denote the subspace of V consisting of functions whose absolute value is Lebesgue measurable, and so Lebesgue integrable. On W let λ be the linear functional of integration with respect to Lebesgue measure m. Then λ, too, is $\mathscr{A}$-invariant so Theorem 12.12 provides a $\mathscr{A}$-invariant linear functional Λ on V that extends λ and is dominated on V by p.

Now for the desired finitely additive measure: if E is a subset of $\mathbb{R}$ with finite outer measure then its characteristic function χ_E is in V (its upper integral is precisely $m^*(E)$), so we can set $\mu(E) = \Lambda(\chi_E)$. The $\mathscr{A}$-invariance of Λ translates into property (12.4) for μ, while the fact that $\Lambda \leq p$ on V shows us that

$$\int_* f = -p(-f) \leq \Lambda(f) \leq p(f) = \int^* f \qquad (f \in V),$$

where on the right we see the *lower integral* of f, i.e., the supremum of the integrals of integrable simple functions that are $\leq f$ at each point of $\mathbb{R}$. In particular, for $f = \chi_E$ with $m^*(E) < \infty$ we obtain (12.5).

It remains only to extend μ to *all* subsets of $\mathbb{R}$, which we do by defining $\mu(E) = \infty$ whenever $m^*(E) = \infty$. With the usual conventions involving arithmetic with ∞, the result is still a finitely additive measure that preserves the desired properties. $\Box$

Higher dimensional extensions? Does the above result extend to affine maps of $\mathbb{R}^N$ for $N > 1$? In this case the maps are $\gamma_{A,v} : x \rightarrow Ax + v$ with A in the group of invertible $n \times n$ real matrices and v a vector in $\mathbb{R}^N$. The change of variable formula now tells us that $\lambda(\gamma_{A,v}(E)) = \det(A)\lambda(E)$ for each Borel subset E of $\mathbb{R}^N$, where λ denotes Lebesgue measure on $\mathbb{R}^N$, so the question is: For $n > 1$ does there exist a finitely additive extension of Lebesgue measure to all the subsets of $\mathbb{R}^N$ that satisfies the above transformation formula.

The answer is "No!" For $N = 3$ the Banach–Tarski Paradox tells us that no such measure exists, even for the subgroup of $A(\mathbb{R}^3)$ consisting of isometries of $\mathbb{R}^3$. The Banach–Tarski Paradox extends to $\mathbb{R}^N$ with $N > 3$ (the proof is an adaptation—not entirely trivial—of the three dimensional one; see, for example, [121, Chap. 5]), with the same result for extensions of Lebesgue measure. For $N = 2$ there is no Banach–Tarski Paradox to help us out here. In its place, however, is the *von Neumann Paradox*, according to which any two bounded subsets of $\mathbb{R}^2$ with nonvoid interior are equidecomposable with respect to the group of affine maps $\gamma_{A,v}$ for which $\det(A) = 1$, i.e., the group of *area-preserving* affine maps. Thus, once again there is no hope for a two dimensional extension of Theorem 12.14.

Countably additive extensions? The usual construction of a subset of $\mathbb{R}$ that's not Lebesgue measurable shows that (assuming the Axiom of Choice) there is no countably additive extension of Lebesgue measure to all subsets of the real line.

12.6 Right vs. Left

Having left the friendly confines of commutativity, we need to address the question of "rightness vs. leftness" for invariant Borel measures on topological groups, and more generally for means on "non-topological" groups (recall Definition 9.16, p. 114). To this point "invariant," for a group G and a mean Λ on $B(G)$ has meant that $L_\gamma^\sharp \Lambda : = \Lambda \circ L_\gamma = \Lambda$ for each of the "left-translation maps" $L_\gamma : B(G) \rightarrow B(G)$ defined for $\gamma \in G$ by

$$L_\gamma(f)(x) = f(\gamma x) \qquad (x \in G, f \in B(G)). \tag{12.6}$$

If a *compact* group G has such an invariant mean (e.g., if G is abelian, or more generally, solvable) then the Riesz Representation Theorem (Sect. 9.2, p. 104) associates with the restriction of this mean to $C(G)$ a similarly invariant regular Borel probability measure—a Haar measure.

For non-commutative groups we need to address the corresponding idea of "right-invariance" that utilizes the transformations $R_\gamma: B(G) \to B(G)$ defined by

$$R_\gamma(f)(x) = f(x\gamma) \qquad (x \in G, f \in B(G)). \tag{12.7}$$

The question arises: "Is right-invariance the same as left invariance?"

Right vs. left Haar measure. We'll see in the next chapter that every compact topological group has a left-invariant regular Borel probability measure; we know right now (Corollary 12.11 above) that such a measure exists if the group is *solvable*. For the time being, however, let's just assume the existence of such a measure for a given compact group and see where this leads.

Theorem 12.15. *Suppose G is a compact topological group and μ is a left-invariant regular Borel probability measure for G. Then:*

(a)   *μ is also right-invariant, hence "bi-invariant."*
(b)   *There is no other invariant regular probability measure for G.*
(c)   *μ is "inversion invariant":*

$$\int f(x^{-1}) d\mu(x) = \int f \, d\mu \qquad (f \in C(G)),$$

i.e., $\mu(B) = \mu(B^{-1})$ for every Borel subset B of G.

Proof. Suppose ν is a right-invariant regular probability measure for G. Then for each $f \in C(G)$ the left-invariance of μ demands that

$$\int f \, d\mu = \int f(xy) d\mu(y) \qquad (x \in G),$$

hence

$$\int f \, d\mu = \int \left(\int f(xy) d\mu(y) \right) d\nu(x) \qquad [\nu(G) = 1]$$

$$= \int \left(\int f(xy) d\nu(x) \right) d\mu(y) \qquad [\text{Fubini}]$$

$$= \int \left(\int f(x) d\nu(x) \right) d\mu(y) \qquad [\nu \text{ right invariant}]$$

$$= \int f \, d\nu \qquad [\mu(G) = 1].$$

Thus $\mu = \nu$, which establishes (a) and (b).

As for (c), note that μ has a natural right-invariant companion $\tilde{\mu}$ defined, thanks to the Riesz Representation Theorem, by

$$\int f \, d\tilde{\mu} = \int f(x^{-1}) d\mu(x) \qquad (f \in C(G)).$$

By (b) we must have $\tilde{\mu} = \mu$, thus establishing the inversion-invariance of μ.   □

Summary: For *compact* topological groups there's no distinction between left- and right-invariant Borel probability measures. Such a "bi-invariant" measure (whose existence we'll prove in the next chapter) is unique, and even "inversion-invariant."

In the next section we'll discuss topological groups that are *not* compact. The following exercise shows that in this generality there may be left-invariant Borel measures that are not right-invariant.

Exercise 12.6 (Haar measure(s) on a non-compact group). Let $G = \{(x,y) \in \mathbb{R}^2 : x > 0\}$ be the open right-half plane of $\mathbb{R}^2$ with the binary operation:

$$(a,b) \cdot (x,y) := (ax, ay + b) \qquad ((a,b),(x,y) \in G).$$

(a)   Show that G, in the operation described above, is a topological group that is *solvable*.
      Suggestion: Examine the map $(x,y) \rightarrow \left[\begin{smallmatrix} x & y \\ 0 & 1 \end{smallmatrix}\right]$ (cf. Example 12.5 and Exercise 12.1).

(b)   Use the change-of-variable formula for double integrals to show that the measure $dx\,dy/x^2$ is left-invariant on G, but not right-invariant.

(c)   Show that the measure $dx\,dy/x$ is right-invariant on G, but not left-invariant.

Right- vs. left-invariant means. For non-abelian groups the situation of left- vs. right-invariance of *means* is more subtle than the one described above for measures. It turns out that left-invariant means need not be right-invariant (and vice versa), but once there is a left- or right-invariant mean, there is a "bi-invariant one." Thus there is no "left vs. right" problem with the notion of "amenable."

In addition to the notions of left and right invariance for means, there is a notion of *inversion-invariance* that mirrors the property observed for invariant measures in Theorem 12.15. Define the linear transformation $J : B(G) \rightarrow B(G)$ by

$$(Jf)(x) = f(x^{-1}) \qquad (f \in B(G),\ x \in G) \tag{12.8}$$

and call a mean Λ on $B(G)$ *inversion invariant* if $J^\sharp \Lambda = \Lambda$, i.e., if $\Lambda \circ J = \Lambda$.

Theorem 12.16. *Suppose G is a group for which $B(G)$ has a left-invariant mean. Then $B(G)$ has a mean that is both bi-invariant and inversion invariant.*

Proof. We'll first show that every left-invariant mean has a right-invariant counterpart. To this end note that for the inversion operator J defined by (12.8),

$$R_\gamma J = J L_{\gamma^{-1}} \qquad \text{and} \qquad L_\gamma J = J R_{\gamma^{-1}}. \tag{12.9}$$

Thus if λ is a left-invariant mean for $B(G)$ then the calculation below shows that $\rho = J^\sharp \lambda$ is right invariant. For every $\gamma \in G$:

$$R_\gamma^\sharp \rho = R_\gamma^\sharp J^\sharp \lambda = (J R_\gamma)^\sharp \lambda = (L_{\gamma^{-1}} J)^\sharp \lambda = J^\sharp L_{\gamma^{-1}}^\sharp \lambda = J^\sharp \lambda = \rho,$$

where the middle equality above comes from the first identity of (12.9) and the next-to-last one from the left-invariance of λ. Clearly $\rho(1) = 1$, and it's easy to check that ρ is a positive linear functional on $B(G)$, hence a right-invariant mean.

From a left-invariant mean λ on $B(G)$ and a right-invariant one ρ, the definition below provides a *bi-invariant* one: a mean v on $B(G)$ with $L_\gamma^\sharp v = R_\gamma^\sharp v = v$ for every $\gamma \in G$:

$$v(f) = \lambda(\tilde{f}) \quad \text{where} \quad \tilde{f}(\gamma) = \rho(L_\gamma f) \qquad (f \in B(G), \gamma \in G).$$

One checks easily that v is a mean on $B(G)$. As for its bi-invariance, a little calculation (using the identity $L_{\gamma\beta} = L_\beta L_\gamma$, and the fact that every left translation commutes with every right translation) shows that for each $\gamma \in G$ and $f \in B(G)$:

$$\widetilde{R_\gamma f} = \tilde{f} \quad \text{and} \quad \widetilde{L_\gamma f} = L_\gamma \tilde{f},$$

whereupon

$$v(R_\gamma f) = \lambda(\widetilde{R_\gamma f}) = \lambda(\tilde{f}) = v(f)$$

and

$$v(L_\gamma f) = \lambda(\widetilde{L_\gamma f}) = \lambda(L_\gamma \tilde{f}) = \lambda(\tilde{f}) = v(f)$$

as desired.

Finally, from the bi-invariant mean v we form an inversion-invariant one $\eta = (v + J^\sharp v)/2$ that is easily to inherit the bi-invariance of v.   □

Example 12.17 (A left-invariant mean that's not right-invariant). Let G be the group of Exercise 12.6; the identity of this group is the point $(1,0)$, and the inverse of $(x,y) \in G$ is $(1/x, -y/x)$.

By part (a) of Exercise 12.6 we know that G is solvable, hence Corollary 12.12, our "solvable" Invariant Hahn–Banach Theorem, applies to $B(G)^\sharp$. In particular, let p denote the gauge function on $B(G)$ defined by the iterated upper limits

$$p(f) = \limsup_{y \to \infty} \left[\limsup_{x \to \infty} f(x,y) \right] \qquad (f \in B(G)).$$

Let W denote the set of all functions $f \in B(G)$ for which the iterated limit

$$\lambda(f) = \lim_{y \to \infty} \left[\lim_{x \to \infty} f(x,y) \right]$$

exists (finitely). One checks easily that:

(a) p is left-invariant on $B(G)$: $p \circ L_\gamma = p$ for every $\gamma \in G$.
(b) W is a linear subspace of $B(G)$ with $L_\gamma(W) \subset W$ for each $\gamma \in G$, and
(c) λ is a linear functional on W that is left-invariant for G.

Since G is solvable, our extended Invariant Hahn–Banach Theorem applies to produce an extension of λ to a left-invariant linear functional Λ on $B(G)$.

However Λ is *not* right-invariant for G. For example, if $g(x,y) := xy/(x^2 + y^2)$ then $g \in W$ with $\lambda(g) = 0$. For $(a,b) \in G$ we have for each $y \in \mathbb{R}$:

$$\lim_{x\to\infty} R_{(a,b)}g(x,y) = \lim_{x\to\infty} \frac{ax(bx+y)}{(ax)^2 + (bx+y)^2} = \frac{ab}{a^2+b^2} = g(a,b).$$

Thus $\lambda(R_{(a,b)}g) = g(a,b)$ which is $\neq 0$ if $b \neq 0$. However $\lambda(g) = 0$, so the functional λ, and therefore its extension Λ, is not right-invariant.

Exercise 12.7. Show that the mean Λ of Example 12.17 is not inversion-invariant.

Exercise 12.8 (Banach limits for solvable groups). Suppose G is an infinite group that is *solvable*. For a function $f : G \to \mathbb{R}$ and $c \in \mathbb{R}$ define "$\lim_{\gamma\to\infty} f(\gamma) = c$" to mean: "For every $\varepsilon > 0$ there exists a finite subset F_ε of G such that $|f(\gamma) - c| < \varepsilon$ for every $\gamma \in G \backslash F_\varepsilon$." Make similar definitions for upper and lower limits. Use Corollary 12.12 to show that there exists a mean Λ on $B(G)$ that is both bi-invariant and inversion-invariant for G, and for which

$$\liminf_{\gamma\to\infty} f(\gamma) \leq \Lambda(f) \leq \limsup_{\gamma\to\infty} f(\gamma)$$

for each $f \in B(G)$.

12.7 The Locally Compact Case

To say a topological space X is *locally compact* means that for each point $x \in X$, every neighborhood of x contains a compact neighborhood of x. In other words, at each point the topology of the space has a local base of compact neighborhoods. As we've mentioned previously (but will not prove here), every locally compact group has a left—and therefore a right—Haar measure. Under appropriate regularity conditions left Haar measure is unique up to positive scalar multiples, as is right Haar measure, but we've already seen (Exercise 12.6) that left and right Haar measures need not be scalar multiples of each other. Detailed proofs of the existence and uniqueness of Haar measure on locally compact groups exist in many places; see, for example, [39, Chap. 2] or [29, Chap. 7]. There does not, however, seem to be a neat functional-analysis proof of this result. The purpose of this section is to show how our "Markov–Kakutani method" can be modified to provide Haar measure, at least for locally compact groups that are *solvable*.

Throughout this discussion it will help to keep in mind three examples: Lebesgue measure on Euclidean space, and the left and right Haar measures on the group G of Exercise 12.6. All three measures are unbounded, and the last two show that, even in the solvable case, left and right Haar measures can be essentially different.

Regular and Radon measures. Suppose μ is a Borel measure for a locally compact (Hausdorff) space X. To say that a Borel set $E \subset X$ is:

– μ-*outer regular* means that $\mu(E) = \inf\{\mu(U) : U$ is open and $U \supset E\}$.
– μ-*inner regular* means $\mu(E) = \sup\{\mu(K) : K$ is compact and $K \subset E\}$.

To say that μ itself is

–   *Regular* means that every Borel set is both inner and outer regular (similarly we can attach to μ the terms "inner regular" or "outer regular").

–   *Locally finite* means that every point of the space has a neighborhood of finite measure.
–   A *Radon measure* means that it is locally finite, outer regular, and every *open set* is μ-inner regular.

We'll see below (Exercise 12.10) that in general not every Radon measure is regular. This is not something to worry about, especially in the compact case:

> *Exercise* 12.9. Show that for *compact* spaces every Radon measure is regular.

Haar Measure To say that μ is a *left Haar measure* for a locally compact group G means that μ is a Radon measure for G that is invariant for left-translation on the group, i.e., $\mu(E) = \mu(gE)$ for every Borel subset E of G and every $g \in G$. Right Haar measure is defined similarly.

We'll need the full-strength version of the Riesz Representation Theorem. If X is a locally compact space that is not compact, the space $C(X)$ of all continuous real-valued functions on X is no longer an appropriate setting for the Riesz theorem; non-compact spaces raise the spectre of unbounded functions and infinite measures, creating problems for the integration of arbitrary continuous functions against arbitrary Borel measures. The resolution is to replace $C(X)$ by its subspace $C_c(X)$: those continuous functions on X that have *compact support*, i.e., that vanish off some compact set. Each such function is bounded and can be integrated against every locally finite Borel measure. Such measures therefore induce linear functionals on $C_c(X)$, the positive ones inducing positive functionals. The Riesz Representation Theorem says that each positive linear functional on $C_c(X)$ is given by integration against such a measure and, with the appropriate conditions of regularity, this representing measure is unique. More precisely (see, e.g., [102, Theorem 2.14, pp. 40–41]):

The Riesz Representation Theorem for locally compact spaces. *Suppose X is a locally compact space and Λ a positive linear functional on $C_c(X)$. Then there is a unique Radon measure for X such that $\Lambda(f) = \int f\,d\mu$ for every $f \in C_c(X)$.*

The exercise below shows that Haar measure—even for a commutative locally compact group—need not always be regular. The group G in question is the additive group $\mathbb{R}^2$ endowed with the product topology it gets when viewed as $\mathbb{R}_d \times \mathbb{R}$, where $\mathbb{R}_d$ denotes the real line with the *discrete* topology.

> *Exercise* 12.10 (A non-regular Haar measure). For the group G described above:
>
> (a)   Show that G is locally compact, and even *metrizable* (for $p_j = (x_j, y_j) \in G(j = 1, 2)$ take $d(p_1, p_2)$ equal to $|y_1 - y_2|$ if $x_1 = x_2$, and $1 + |y_1 - y_2|$ otherwise).
>
> (b)   Let δ denote the counting measure for $\mathbb{R}_d$ and let λ denote Lebesgue measure on the Borel subsets of $\mathbb{R}$. Show that each of these is a Haar measure for its respective topological group.
>
> (c)   Show that $\mathbb{R}_d \times \{0\}$ (the "discrete x-axis" has μ-measure ∞, whereas each of its compact subsets has μ-measure zero. Thus μ is not regular. Show that μ is, nevertheless, a Radon measure.
>
> (d)   Show that μ is a Haar measure for G.

Here is the main result of this section.

Theorem 12.18. *Every locally compact solvable topological group has a Haar measure.*

Strategy of proof. Assume G is solvable and locally compact, but not compact. Let e denote the identity element of G. We seek to produce left Haar measure for G by applying our "solvable" Markov–Kakutani theorem to an appropriate subset of the algebraic dual $C_c(G)^\sharp$ of $C_c(G)$. Since non-zero constant functions no longer belong to $C_c(G)$, the space of "means" that worked so well in the compact situation no longer exists. Our argument will hinge on finding a substitute.

Simplifying assumptions. To keep the argument as transparent as possible we'll assume G is *metrizable* and that its metric d is *G-invariant* in the sense that $d(\gamma x, \gamma y) = d(x, y)$ for all $x, y, \gamma \in G$ (i.e., for each $\gamma \in G$ the left-translation map $L_\gamma : x \to \gamma x$ ($x \in G$) is an isometry). See the *Notes* at the end of this chapter for some discussion of these assumptions.

Notation. Let $B_r(x)$ denote the open d-ball of radius $r > 0$, centered at $x \in G$. The G-invariance of d insures that $\gamma B_r(x) = B_r(\gamma x)$ for all $x, \gamma \in G$. For $f \in C_c(G)$ we'll define $\|f\| = \max\{|f(x)| : x \in G\}$, where compactness of support insures the (finite) existence of the maximum.

Small and large functions. Since G is locally compact it has, at each point, a base of compact neighborhoods. In particular, there exists $r > 0$ such that $B_r(e)$ has compact closure. Thus $B_r(x) = xB_r(e)$ has compact closure for each $x \in G$. Fix this radius r for the rest of the proof.

Let $C_c^+(G)$ denote the collection of non-negative functions in $C_c(G)$. To say that $f \in C_c^+(G)$ is:

-   *Small* means that its values are all ≤ 1 on G and its support lies in $B_{r/2}(x)$ for some $x \in G$.
-   *Large* means that its values are all ≥ 1 on $B_r(x)$ for some $x \in G$.

Quasimeans. We'll call a positive linear functional on $C_c(G)$ a *quasimean* if it takes values ≤ 1 on small functions in $C_c^+(G)$ and ≥ 1 on large ones. Let $\mathcal{Q}$ denote the collection of quasimeans. We'll prove the existence of Haar measure for G by showing that $\mathcal{Q}$ is a nonempty, convex, weak-star compact subset of $C_c(G)^\sharp$ that is taken into itself by each translation-adjoint $L_\gamma^\sharp$. The usual argument involving the (solvable) Markov–Kakutani Theorem and the Riesz Representation Theorem will then lead to the desired Haar measure.

It's easy to check that $\mathcal{Q}$ is convex and, thanks to the fact that $\gamma B_r(x) = B_r(\gamma x)$, is also invariant under $L_\gamma^\sharp$ for each $\gamma \in G$.

$\mathcal{Q}$ is weak-star closed. Suppose $\Lambda \in C_c(G)^\sharp$ is a weak-star limit point of $\mathcal{Q}$. We wish to show that $\Lambda \in \mathcal{Q}$. For each $\varepsilon > 0$ and finite subset F of $C_c(G)$, the weak-star neighborhood of Λ

$$N(\Lambda, F, \varepsilon) = \{\Gamma \in C_c(G)^\sharp : |\Gamma(f) - \Lambda(f)| < \varepsilon \ \forall f \in F\}$$

contains a point of $\mathcal{Q}$. Suppose, then, that $f, g \in C_c^+(G)$ with f small and g large. Fix $\varepsilon > 0$, and choose $\Gamma \in N(\Lambda, \{f, g\}, \varepsilon) \cap \mathcal{Q}$. Then

$$\Lambda(f) \leq \Gamma(f) + \varepsilon \leq 1 + \varepsilon \qquad \text{and} \qquad \Lambda(g) \geq \Gamma(g) - \varepsilon \geq 1 - \varepsilon.$$

Since ε is an arbitrary positive number, $\Lambda(f) \leq 1$ and $\Lambda(g) \geq 1$, hence $\Lambda \in \mathcal{Q}$, as desired.

$\mathcal{Q}$ *is weak-star compact.* Since we now know that $\mathcal{Q}$ is weak-star closed, to show it's compact we need only prove that it's pointwise bounded on $C_c(G)$ (Corollary 9.15, p. 112). To this end fix $\Lambda \in \mathcal{G}$ and note that, thanks to the definition of "small" function: for every $f \in C_c^+(G)$ with support contained in some ball of radius $r/2$ we have $\Lambda(f) \leq \|f\|$. Now for arbitrary $f \in C_c^+(G)$ we can cover its (compact) support by a finite number of open d-balls of radius $r/2$. Lemma B.6 (p. 190) provides a partition of unity

$\{p_1, p_2, \ldots, p_n\}$ subordinate to that cover. Thus $f = \sum_{j=1}^n p_j f$ where each function $p_j f$ belongs to $C_c^+(G)$ and has support contained in a ball of radius $r/2$. It follows that

$$\Lambda(f) = \sum_{j=1}^n \Lambda(p_j f) \leq \sum_{j=1}^n \|p_j f\| \leq n \|f\|,$$

where the integer n depends on f, but not on Λ. Thus $\mathcal{Q}$ is pointwise bounded on $C_c^+(G)$.

Now suppose $f \in C_c(G)$. Then $f = f_+ - f_-$, the difference of two functions in $C_c^+(G)$, each of which has norm $\leq \|f\|$ and support contained in that of f. Thus

$$|\Lambda(f)| \leq \Lambda(f_+) + \Lambda(f_-) \leq n\|f_+\| + n\|f_-\| \leq 2n\|f\|,$$

where n does not depend on Λ. Thus $\mathcal{Q}$ is pointwise bounded on $C_c(G)$, hence weak-star compact in $C_c(G)^{\sharp}$.

$\mathcal{Q}$ *is nonempty.* For most proofs this sort of statement is a triviality. Not so here! None of the "usual suspects" (the point evaluations) belong to $\mathcal{Q}$. (Exercise: Why not?) What's needed is a subset S of G having the following properties:

(S1) S has at least one point in each open d-ball of radius r, and
(S2) S has no more than one point in each open ball of radius $r/2$.

Example 12.19. $G = \mathbb{R}$ and $r = 1$. Then $S = \mathbb{Z}$ has the desired properties. Note: S is *maximal* with respect to the property that any pair of its distinct elements lies at least 1 unit apart.

We're going to show that such a set S exists in every group G of the sort we're considering. Assuming this for the moment, define the functional Λ on $C_c(G)$ by

$$\Lambda(f) := \sum_{s \in S} f(s) \qquad (f \in C_c(G)). \tag{12.10}$$

Since we can cover each compact subset of G by finitely many open balls of radius $r/2$, each such set can contain at most finitely many points of S, hence for each

$f \in C_c(G)$ only finitely many summands on the right-hand side of (12.10) are non-zero. The right-hand side of (12.10) therefore makes sense and provides a positive linear functional on $C_c(G)$.

Suppose $f \in C_c^+(G)$ is "small," i.e., has support in a ball of radius $r/2$ and all values ≤ 1. Each such ball contains no more than one point of S, so $\Lambda(f) = 0$ or 1. Thus $\Lambda(f) \leq 1$ on "small" functions in $C_c^+(G)$. Suppose on the other hand that $f \in C_c^+(G)$ is "large," i.e., takes only values ≥ 1 on some ball of radius r. Then $\Lambda(f) \geq 1$ since this ball must contain a point of S. Thus $\Lambda \in \mathscr{Q}$, proving that $\mathscr{Q}$ is not empty.

The proof that G harbors the desired set S is inspired by Example 12.19.

Claim. *Suppose $S \subset G$ is maximal with respect to the property*

$$s,t \in S \text{ with } s \neq t \implies d(s,t) \geq r. \tag{*}$$

Then S satisfies conditions (S1) and (S2) above.

Proof of Claim. To check S has property (S1), note that if this were not the case there would be a point $x \in G$ at distance $\geq r$ from each point of S. Then $S \cup \{x\}$, which properly contains S, would obey (*) thus contradicting the maximality of S. As for (S2), suppose $s,t \in S$ lie in the ball $B_{r/2}(x)$. Then by the triangle inequality $d(s,t) < r$, hence $s = t$, thus establishing the Claim.

It remains to prove the existence of our maximal S. Let $\mathscr{T}$ denote the family of subsets T of G with the property (*).

$\mathscr{T}$ is nonempty. Since we're assuming the closure of $B_r(e)$ is compact, while G is not, there must exist $x \in G$ with $d(e,x) > r$. Thus $\{e,x\} \in \mathscr{T}$, so the family $\mathscr{T}$ is nonempty.

Enter Zorn's Lemma. If $\mathscr{C}$ is a subfamily of $\mathscr{T}$ that is *totally ordered* by inclusion (i.e., given two members of $\mathscr{C}$, one of them is contained in the other), then the union of the sets that are elements of $\mathscr{C}$ belongs to $\mathscr{T}$ (exercise). Thus each subfamily of $\mathscr{T}$ that is totally ordered has an upper bound, so by Zorn's Lemma (Appendix E.3) $\mathscr{T}$ has a maximal element S (note that this argument used only the fact that G is a metric space and $r < \sup_{x,y \in G} d(x,y)$).

Concluding the proof. We now have $\mathscr{Q}$, our nonempty, convex, weak-star compact subset of $C_c(G)^\sharp$, and the family $\mathscr{L}^\sharp$ of continuous affine (in fact linear) self-maps $L_\gamma^\sharp$ of $\mathscr{Q}$ ($\gamma \in G$). The argument in the paragraph preceding Theorem 12.6 shows that the family of maps $\mathscr{L}^\sharp$ inherits the solvability of G, hence our extended Markov–Kakutani Theorem (Theorem 12.1) guarantees that $\mathscr{L}^\sharp$ has a fixed point Λ in $\mathscr{Q}$. The measure provided for Λ by Riesz Representation Theorem is, by a familiar argument, the Haar measure we seek. □

Notes

Solvable families of maps. The original source for this is Day's paper [27, p. 285]; see also [37, Theorem 3.2.1, pp. 155–156].

The symmetric group. Exercise 12.4 is from [108, p. 253]. The fact that S_n is not solvable for $n \geq 5$ is a crucial step in the proof of Abel's Theorem: *For each $n \geq 5$ there is a polynomial of degree n whose roots can not all be found by radicals.* See, for example, Hadlock's Carus Monograph [43, Chap. 3].

Corollary 12.8. The amenability of solvable groups is due to von Neumann [88].

Right vs. left. Theorem 12.16, showing that each left-invariant mean gives rise to a bi-invariant one, is due to M.M. Day [27, Lemma 7, p. 285].

von Neumann's Paradox. The original source is [88]. For a modern exposition in English, see [121, Theorem 7.3, p. 99].

Exercise 12.10. This is taken directly from [102, Chap. 2, Exercise 17, p. 59].

Haar measure for solvable locally compact groups: Those simplifying assumptions. Every metrizable topological group has an invariant metric. In fact the *Birkhoff– Kakutani Theorem* asserts that every *first countable* group is metrizable and has such a metric. See [29, Corollary 3.10, p. 53] or [80, Sect. 1.22, pp. 34–36] for a proof, and [13, 57] for the original papers. With a bit more care the entire proof given above for the existence of Haar measure can be carried out for *every* solvable locally compact group. See Izzo [54] for how to do this in the commutative case; the solvable one being no different. The argument given above is just a translation of Izzo's proof to the solvable, invariantly metrizable case.

Chapter 13
Beyond Markov–Kakutani

THE RYLL–NARDZEWSKI FIXED-POINT THEOREM

Overview. In the last chapter we extended the Markov–Kakutani Theorem—originally proved only for *commuting* families of continuous affine maps—to "solvable" families of such maps. We used our enhanced theorem to show that every solvable group is amenable and that Haar measure exists for every topological group that is both solvable and compact. By contrast, we've seen (Chap. 11) that the group $\mathscr{R}$ of origin-centric rotations of $\mathbb{R}^3$ is paradoxical, hence not amenable, and therefore not solvable. Now $\mathscr{R}$ is naturally isomorphic to the group SO(3) of 3×3 orthogonal real matrices with determinant 1 (Appendix D), a group easily seen to be compact. Thus not every compact group is amenable.

Conclusion: Fixed-point theorems that produce invariant means cannot prove the existence of Haar measure for every compact group.

In this chapter we'll turn to a fixed-point theorem in which the Markov–Kakutani hypothesis of solvability is replaced by a topological condition of "uniform injectivity." This result, due to the Polish mathematician Czesław Ryll–Nardzewski, works with an appropriate modification of our previous duality method to provide Haar measure for *all* compact topological groups. For ease of exposition we'll focus on compact groups that are *metrizable,* sketching afterwards how to make the arguments work in general. Finally, we'll identify the Haar measure for SO(3).

Throughout this chapter we'll be working in vector spaces over the real numbers.

13.1 Introduction to the Ryll–Nardzewski Theorem

Theorem 13.1 (Ryll–Nardzewski [105]). *Suppose X is a* locally convex *topological vector space in which K is a nonvoid, compact, convex subset. Suppose $\mathscr{S}$ is a* semigroup *of continuous, affine self-maps of K that is* uniformly injective. *Then $\mathscr{S}$ has a fixed point in K.*

© Springer International Publishing Switzerland 2016
J.H. Shapiro, *A Fixed-Point Farrago*, Universitext,
DOI 10.1007/978-3-319-27978-7_13

We'll first seek to understand the hypotheses of the Ryll–Nardzewski Theorem after which we'll prove it for the *duals of separable Banach spaces,* a special case that's still general enough to provide the existence of Haar measure for compact, metrizable groups, and which gives an accurate guide to the proof of the general theorem. Let's start with the undefined terms in the Theorem's statement, taking them in the order in which they occur.

Semigroup. A set with an associative binary operation. If $\mathscr{F}$ is a family of self-maps of some set S, then the *semigroup generated by $\mathscr{F}$* (its operation being composition of maps) consists of all possible finite compositions of maps in $\mathscr{F}$. This is the smallest semigroup of self-maps of S containing $\mathscr{F}$; it has the same set of common fixed points as $\mathscr{F}$, and one can even throw in the identity map without changing the common-fixed-point set. Thus when considering fixed points for families of self-maps one need only consider "compositional semigroups with identity."

Locally Convex. For a topological vector space (always assumed Hausdorff) this property means that each point has a base of convex neighborhoods. We have already worked with several important examples of locally convex spaces:

(a)   Normed linear spaces: the balls centered at a given point are convex and form a base for the neighborhoods of that point.

(b)   The space $\mathbb{R}^S$ of all real-valued functions on a set S, with its topology of "pointwise convergence": the basic neighborhoods $N(f, F, \varepsilon)$ for this topology, as defined by Eq. (9.9) (p. 110), are all convex.

(c)   The weak-star topology induced on the algebraic dual $V^\sharp$ of a real or complex vector space V.

A version of the Hahn–Banach Theorem guarantees, for each locally convex topological vector space, the existence of enough continuous linear functionals to separate distinct points of the space, and more generally, to separate disjoint closed convex sets.[1] The exercise below shows that in the absence of local convexity such separation is not guaranteed.

> *Exercise 13.1* (Non locally convex pathology[2]). For $0 < p < 1$ consider the space $L^p = L^p([0,1])$ consisting of (a.e.-equivalence classes of) real-valued Lebesgue measurable functions f on the unit interval for which $\|f\| = \int_0^1 |f(x)|^p \, dx < \infty$ (omission of the p-th root of the integral on the right is deliberate). For $f, g \in L^p$ let $d(f, g) = \|f - g\|$. Show that:
>
> (a) $\|\cdot\|$ is not a norm, but d is a metric making L^p into a topological vector space.
>
> (b)   On L^p the topology induced by the metric d is not locally convex. In fact, the only open (nonempty) convex set is the whole space!
>
> (c)   The only continuous linear functional on L^p is the zero-functional.

Uniformly Injective. "Injective" is another way of saying "one-to-one." To say a family of maps $\mathscr{F}$ taking a set S into a topological vector space X is *uniformly*

---

[1] See, e.g., [103, Chap. 3, pp. 56–62].

[2] For more details, see [103, Sect. 1.47, pp. 36–37].

injective[3] means that for each pair of distinct points $s, t \in S$ the zero vector does not belong to the closure of the set $\{f(s) - f(t) : f \in \mathscr{F}\}$. If X is a normed linear space, then uniform injectivity for $\mathscr{F}$ means that for every pair $s, t \in S$ with $s \neq t$ there exists a positive number $\delta = \delta(s, t)$ such that:

$$\delta < \|f(s) - f(t)\| \quad \text{for every} \quad f \in \mathscr{F}. \tag{13.1}$$

Why Injectivity? Let $X = \mathbb{R}$, $K = [0, 1]$, and consider the two-element compositional semigroup $\mathscr{S} = \{\varphi, \psi\}$, where $\varphi \equiv 0$ and $\psi \equiv 1$. Thus X is locally convex, K is a nonempty, compact, convex subset of X, and $\mathscr{S}$ is a finite semigroup of affine, continuous self-maps of K that does not have a common fixed point. The following prototype of the Ryll–Nardzewski Theorem shows that the culprit here is "lack of injectivity."

Proposition 13.2. *Suppose K is a nonvoid, compact, convex subset of a topological vector space. Then every finite semigroup of continuous, injective, affine self-maps of K has a common fixed point.*

Proof. Let $\mathscr{S}$ denote our finite semigroup of maps. As noted earlier, there is no loss of generality in assuming that it contains the identity map e_K on K.

Claim. $\mathscr{S}$ is a group.

Proof of Claim. We need only show that each map in $\mathscr{S}$ has an inverse. Fix $A \in \mathscr{S}$ and note that since $\mathscr{S}$ is finite there exist positive integers n and m with $1 \leq m < n$ such that $A^n = A^m$, (where, e.g., A^n denotes the composition of A with itself n times). Thus $A^n = A^m A^{n-m} = A^n A^{n-m}$, and since A^n is injective

$$e_K = A^{n-m} = A^{n-m-1}A = AA^{n-m-1}.$$

This exhibits A^{n-m-1} (which exists and belongs to $\mathscr{S}$ because $n - m \geq 1$) as the compositional inverse of A, thus proving the Claim.

Having established that $\mathscr{S}$ is a group, let $A_1, A_2, \ldots, A_n$ denote its elements, and denote by A_0 the arithmetic mean of these elements:

$$A_0 x = \frac{1}{n} \sum_{j=1}^{n} A_j x \qquad (x \in K).$$

Now A_0, though perhaps not a member of $\mathscr{S}$, is nonetheless a continuous, affine self-map of K. The Markov–Kakutani Theorem[4] therefore guarantees that A_0 has a fixed point x_0 in K.

---

[3] Alternative terminology: "non-contracting," or in dynamical systems: "distal."

[4] For this we need only the "single-map" version: Proposition 9.8, p. 108.

Fix $A \in \mathcal{S}$. Since A is affine it respects convex combinations, so $AA_0 = (1/n)\sum_{j=1}^{n} AA_j$. Since $\mathcal{S}$ is a group, the n-tuple $(AA_1, AA_2, \ldots, AA_n)$ is a permutation of the original list $(A_1, A_2, \ldots, A_n)$ of the elements of $\mathcal{S}$. Conclusion: $AA_0 = A_0$. Consequently

$$x_0 = A_0 x_0 = AA_0 x_0 = A x_0$$

i.e., x_0 is a fixed point for A, hence a common fixed point for $\mathcal{S}$. $\square$

On the other hand, for *infinite* semigroups of affine, continuous maps: *injectivity alone is not enough to guarantee a common fixed point*. Once again let $X = \mathbb{R}$ and $K = [0,1]$, but now consider the (infinite) semigroup $\mathcal{S}$ generated by the pair of injective affine self-maps $\varphi(x) = (2x+1)/4$ and $\psi(x) = (x+1)/2$ of K. Since φ and ψ have no common fixed point, neither does $\mathcal{S}$. The exercise below shows what's wrong.

Exercise 13.2. Show that $\mathcal{S}$ as described above is not *uniformly* injective.

13.2 Extreme points of convex sets

For the proof of the Ryll–Nardzewski Theorem we'll make frequent use of the concepts of convex set, convex combination, and convex hull, as set out in Appendix C, Sect. C.1. Here's a crucial addition to this list.

Definition 13.3 (Extreme point). For a convex subset C of a real vector space, an *extreme point* is a point of C that does not lie in the interior of the line segment joining two distinct points of C (i.e., a point that cannot be written as $tx + (1-t)y$, with $0 < t < 1$ and x, y distinct points of C).

Examples of extreme points. The endpoints of a closed interval of the real line, the vertices of a triangle in $\mathbb{R}^2$, or more generally a convex polygon in $\mathbb{R}^N$ (e.g., the standard simplex Π_N). Every point on the boundary of a closed ball in $\mathbb{R}^N$.

Non-examples. In a normed space: each point in the interior of a closed ball. For a convex polygon in $\mathbb{R}^N$: each point that is not a vertex (e.g., for Π_N, each point that is not one of the standard basis vectors for $\mathbb{R}^N$).

Exercise 13.3. Suppose C is a convex subset of a real vector space. Then $p \in C$ is an extreme point if and only if p cannot be represented nontrivially as a convex combination of other points of C.

A key step in our proof of the Ryll–Nardzewski Theorem will involve the following fundamental result about extreme points. If S is a subset of a topological vector space, we'll use the notation $\overline{\mathrm{conv}} S$ for the closure of its convex hull.

Theorem 13.4 (A Krein–Milman theorem). *Suppose K_0 is a nonempty compact subset of a locally convex topological vector space X and that $K = \overline{\mathrm{conv}}\, K_0$ is also compact.[5] Then K_0 contains an extreme point of K.*

This result is a consequence of two famous theorems about nonempty compact subsets K of locally convex spaces. First there is *The* Krein–Milman Theorem, which asserts that not only does K *have* extreme points, it is in fact the *closed convex hull* of these extreme points. Next, the "Milman Inversion" of this theorem says that if K is the closed-convex hull of a compact subset K_0, then all of K's extreme points belong to K_0.[6] For our purposes we'll only need a special case of Theorem 13.4 (Theorem 13.6 below).

We begin with an even more special case of Theorem 13.4.

Lemma 13.5. *Suppose K_0 is a nonempty compact subset of an inner-product space. If $K := \overline{\mathrm{conv}}\, K$ is compact then some point of K_0 is an extreme point of K.*

Proof. Let's denote the ambient inner-product space by X, its inner product by $\langle \cdot, \cdot \rangle$, and its norm by $\| \cdot \|$ (i.e., $\|x\|^2 = \langle x, x \rangle$ for each $x \in X$). Since K_0 is compact there is a smallest closed ball B in X that contains it, and so also contains the closure of its convex hull. Upon making an appropriate translation and dilation we may without loss of generality assume that B is the closed unit ball of X. The compactness of K_0 insures that it intersects ∂B in some vector v. This unit vector (or, for that matter, every unit vector in K_0) will turn out to be the desired extreme point for K. This is obvious from a picture; for an analytic proof suppose $v = tx + (1-t)y$ for some vectors $x, y \in K$ and for some $0 < t < 1$. Since $\|x\|$ and $\|y\|$ are both ≤ 1,

$$1 = \|v\|^2 = \langle v, v \rangle = \langle v, tx + (1-t)y \rangle = t\langle v, x \rangle + (1-t)\langle v, y \rangle$$

$$\leq t\|v\|\|x\| + (1-t)\|v\|\|y\| \leq 1,$$

where the next-to-last inequality follows from the Cauchy–Schwarz Inequality applied to both $\langle v, x \rangle$ and $\langle v, y \rangle$. Thus there is equality throughout, in particular

$$\|x\| = \|y\| = \langle v, x \rangle = \langle v, y \rangle = 1.$$

By the case of equality in the Cauchy–Schwarz inequality, this requires $x = \pm v$ and $y = \pm v$. They can't both be $-v$ lest $v = -v$, i.e., $v = 0$, contradicting the fact that v is a unit vector. On the other hand if one of them is v and the other is $-v$, then $v = \pm(2t - 1)v$ whereupon t is either 0 or 1, another contradiction. Thus v is an extreme point of K. □

The version of Theorem 13.4 that we'll actually need is the following consequence of Lemma 13.5. Recall from Sect. 9.5 that if Y is a real vector space then

---

[5] Compactness of the closed convex hull of a compact set is automatic for Banach spaces (Proposition C.6, p. 195), but not so in general; it can fail even for non-closed subspaces of Hilbert space (Remark C.7, p. 195).

[6] For proofs of these theorems see, e.g., [103], Theorems 3.23 and 3.25, respectively, pp. 75–77.

the *weak-star topology* on the algebraic dual $Y^{\sharp}$ of Y is just the restriction to $Y^{\sharp}$ of the product topology of $\mathbb{R}^Y$. If Y is a *topological* vector space then its *dual space* Y^* is the collection of linear functionals on Y that are *continuous*. Note that Y^* is a linear subspace of $Y^{\sharp}$, and its weak-star topology is just the restriction of the weak-star topology of $Y^{\sharp}$, i.e., the topology of pointwise convergence on Y. We call Y the *predual* of Y^*.

Theorem 13.6 (A Krein–Milman Theorem for separable preduals). *Suppose X is the dual of a separable Banach space, and $K_0 \subset X$ is nonempty and weak-star compact. If $\overline{\mathrm{conv}}\, K_0$ is weak-star compact, then it has an extreme point that lies in K_0.*

Proof. We are assuming that $X = Y^*$, where Y is a separable (real) Banach space. The closed unit ball Y_1 of Y has a countable dense subset $\{y_n\}_1^{\infty}$ (exercise). Each element of $f \in X$, being a continuous linear functional on Y, is bounded on Y_1. Consequently the formula

$$\langle f, g \rangle = \sum_{n=1}^{\infty} \frac{1}{2^n} f(y_n) g(y_n) \qquad (f, g \in X) \tag{13.2}$$

makes sense, and defines a bilinear form on X that is, in fact, an *inner product*. To see why, define $\|f\| := \sqrt{\langle f, f \rangle}$ for $f \in X$. To say that $\langle \cdot, \cdot \rangle$ is an inner product is to say that the seminorm $\| \cdot \|$ is a norm, i.e., that $\|f\| = 0$ only when $f = 0$. If $\|f\| = 0$ then by (13.2) we have $f(y_n) = 0$ for $n = 0, 1, 2, \ldots$, hence $f = 0$ on Y_1 by the continuity of f on Y and the density of $\{y_n\}_0^{\infty}$ in Y_1. Since f is a linear functional it must therefore vanish on all of Y.

CLAIM. *The norm topology η induced on X by this inner product coincides on weak-star compact sets with the weak-star topology ω.*

Once we've proved this Claim, the desired result on extreme points will follow from Lemma 13.5.

Proof of Claim. Let K be a weak-star compact subset of X. We need only show that the topology η is weaker than ω. Once this is done we'll know that the identity map j on K is continuous from ω to η, and so takes ω-closed subsets of K (which are ω-compact) to η-compact subsets of K (which are η-closed). Thus j is not just a continuous map from (K, ω) to (K, η), but also a *closed* one, and so (upon taking complements) an *open* one. Thus j is a homeomorphism, i.e., $\omega = \eta$.

To show that η is weaker than ω, note that by Proposition 9.10 (p. 111) we know that each vector $y \in Y_1$ induces an ω-continuous function $\hat{y} \colon K \to \mathbb{R}$ via the definition $\hat{y}(f) = f(y)$ $(f \in K)$. Since K is ω-compact, $\hat{y}$ is bounded thereon for each $y \in Y_1$. Turning things around: K is *pointwise bounded* on Y_1, hence by the Uniform Boundedness Principal[7] there exists a positive number M such that $|f| < M$ on Y_1 for every $f \in K$.

Now fix $\varepsilon > 0$ and a point $f_0 \in K$, and consider the relatively open subset U of K obtained by intersecting the η-ball of radius ε and center f_0 with K. Choose

---

[7] See, e.g., [102, Theorem 5.8, pp. 98–99], where it's called the "Banach-Steinhaus Principle."

a positive integer N for which $\sum_{n=N+1}^{\infty} 2^{-n} < \varepsilon^2/(8M^2)$. Let $F = \{y_0, y_1, \ldots, y_N\}$ be a finite subset of Y and suppose $f \in N(f_0, F, \varepsilon/\sqrt{2})$. Then, continuing with the notation $\|\cdot\|$ for the norm induced by the inner product (13.2):

$$\|f - f_0\|^2 = \sum_{n=1}^{\infty} \frac{1}{2^n} |f(y_n) - f_0(y_n)|^2 = \sum_{n=1}^{N} + \sum_{n=N+1}^{\infty} \frac{1}{2^n} |f(y_n) - f_0(y_n)|^2$$

$$< \frac{\varepsilon^2}{2} \sum_{n=1}^{N} \frac{1}{2^n} + (2M)^2 \sum_{n=N+1}^{\infty} \frac{1}{2^n} < \frac{\varepsilon^2}{2} + \varepsilon^2 \frac{4M^2}{8M^2}$$

$$= \frac{\varepsilon^2}{2} + \frac{\varepsilon^2}{2} = \varepsilon^2.$$

Thus $N(f_0, F, \varepsilon/\sqrt{2}) \subset U$. We've shown that if $f_0 \in K$ then every η-neighborhood of f_0 contains an ω neighborhood of f_0, i.e., that the topology η induced on X by the inner product (13.2) is weaker than—and therefore equal to—the weak-star topology induced on X by its predual Y.   □

The metrizability argument given above produces the following useful result:

Proposition 13.7 (Weak-star metrizability). *If X is the dual of a separable Banach space, then on each compact subset of X the weak-star topology is metrizable.*

In the next section we'll prove our "separable-predual" version of the Ryll–Nardzewski Theorem. The action will take place in the dual space of $C(G)$, where G is a compact, metrizable group, so we will need to know that $C(G)$ is separable. According to Proposition B.7 of Appendix B, this is true even if G is just a compact metric space.

13.3 Ryll–Nardzewski: separable predual version

In this section, X will be the (topological) dual of a separable (real) Banach space Y. Instead of considering X in its norm topology, however, we will endow it with the weak-star topology it gets from its predual Y. Thus by Proposition 13.7, every (weak-star) compact subset of X will be *metrizable*.

Theorem 13.8 ("Ryll–Nardzewski lite"). *Suppose X is the dual of a separable Banach space, K is a nonempty convex, weak-star compact subset of X, and $\mathscr{S}$ is a uniformly injective semigroup of continuous, affine self-maps of K. Then $\mathscr{S}$ has a common fixed point in K.*

Proof. The argument is best broken into several pieces.

Step I. It is enough to show:

(*)   *Every finite subset of $\mathscr{S}$ has a common fixed point.*

For suppose we've established (*). If $A \in \mathscr{S}$ let F_A denote the fixed-point set of A:

$$F_A = \{x \in K : Ax = x\}.$$

We wish to show that $\bigcap \{F_A : A \in \mathscr{S}\}$ is nonempty. By the continuity of each map in $\mathscr{S}$ we know that each fixed-point set F_A is closed in K. Now (*) is the assertion that the family of all these sets has the finite intersection property. Thus the compactness of K insures the entire family has nonvoid intersection.

Step II. Fix a finite subset $\mathscr{A} = \{A_1, A_2, \dots A_n\}$ of $\mathscr{S}$. By Step I we'll be done if we can show that $\mathscr{A}$ has a common fixed point, so as noted above (Sect. 13.1, p. 164), we may as well assume that $\mathscr{S}$ is the semigroup generated by $\mathscr{A}$. Note that even though it is finitely generated, $\mathscr{S}$ need not be finite—if it were, we'd be done by Proposition 13.2. Nevertheless, as in the proof of that Proposition, we'll pin our hopes on the affine continuous self-map $A_0 = (A_1 + A_2 + \cdots + A_n)/n$ of K, for which the Markov–Kakutani Theorem once again guarantees a fixed point $x_0 \in K$. As in the case of finite $\mathscr{S}$, we'll show that x_0 is also a fixed point for each of the maps $A_1, \dots, A_n$. Now, however, our argument needs to be more subtle.

Step III. Let $\mathscr{S}x_0 = \{Ax_0 : A \in \mathscr{S}\}$: the $\mathscr{S}$-orbit of x_0, and consider its closure K_0, a compact subset of K. Since $C = \overline{\mathrm{conv}}\, K_0$ is a closed subset of K, it too is compact, so by Theorem 13.6 some point e of K_0 is an extreme point of C. Since K is metrizable (Proposition 13.7) there is a sequence (T_j) of maps in $\mathscr{S}$ such that $T_j x_0 \to e$. Since $A_0 x_0 = x_0$ we have

$$e = \lim_j (T_j A_0) x_0 = \lim_j T_j \left(\frac{A_1 x_0 + A_2 x_0 + \cdots + A_n x_0}{n} \right)$$

$$= \lim_j \left(\frac{(T_j A_1) x_0 + (T_j A_2) x_0 + \cdots + (T_j A_n) x_0}{n} \right).$$

In the last line, which follows from the affine-ness of the maps T_j, we're looking at n sequences $((T_j A_k) x_0)_{j=1}^{\infty}$ for $k = 1, 2, \dots, n$, each drawn from $\mathscr{S}x_0$. Thanks to the weak-star compactness and metrizability of K we can find a single subsequence (j_i) of indices such that the sequence $((T_{j_i} A_k) x_0)_{i=1}^{\infty}$ converges for each k to a vector $y_k \in K_0 = \overline{\mathscr{S}x_0}$. Thus the vector e, which we know belongs to $C = \overline{\mathrm{conv}}\, K_0$, is actually the average of the vectors $y_1, y_2, \dots, y_n \in K_0$ and so belongs to $\mathrm{conv}\, K_0$. Since e is an extreme point of $\overline{\mathrm{conv}}\, K_0$ the y_k's must all be equal to e.

Step IV. Recall that we're trying to show that $x_0 = A_k x_0$ for each $1 \le k \le n$. Since $A_0 x_0 = x_0$ it's enough to know (definition of A_0) that all the vectors $A_k x_0$ are the same. Choose two of them, say $A_\mu x_0$ and $A_\nu x_0$. We know from Step III that

$$0 = e - e = y_\mu - y_\nu = \lim_i [T_{j_i}(A_\mu x_0) - T_{j_i}(A_\nu x_0)],$$

so the zero vector belongs to the closure of the set $\{T(A_\mu x_0) - T(A_\nu x_0) : T \in \mathscr{S}\}$. This, plus the uniform injectivity of $\mathscr{S}$ (at last!), guarantees that $A_\mu x_0 = A_\nu x_0$. $\qquad\square$

Ryll–Nardzewski at full strength (sketch of proof). To upgrade the proof just presented to one that establishes Theorem 13.1:

(a)   In Step III: Instead of Theorem 13.6, the "separable predual" version of "A Krein–Milman Theorem," use the full-strength one, Theorem 13.4.

(b)   In Step IV: Instead of sequences and subsequences, use "nets" and "subnets" (see the *Notes* below for the definition and discussion of these concepts). The proof just given then goes through *mutatis mutandis*. The reason for avoiding this generality is that, while nets provide a straightforward generalization of sequences, the same cannot be said for subnets vs. subsequences. In fact subnets of sequences need not be subsequences [123, Problem 11B, p. 77].

13.4 Application to Haar Measure

In the Ryll–Nardzewski Theorem we finally have a result that allows the duality method of Chaps. 9, 10, and 12 to establish the existence of Haar measure for *every* compact topological group—commutative or not. Recall the basics of this method: To each element γ of the group G we assign the left-translation operator L_γ defined on either $C(G)$ (the space of real-valued functions on G that are *continuous*) or $B(G)$ (the space of real-valued functions on G that are *bounded*) by $L_\gamma f(x) = f(\gamma x)$ ($x \in G$). We implore a kind spirit to grant us a common fixed point for the collection of algebraic adjoints of these operators. The Riesz Representation Theorem then transforms this fixed point into Haar measure for G.

In Chap. 9 the group G was commutative and our translation adjoints lived on the algebraic dual of $C(G)$. In Chap. 10 we observed that the same argument worked as well in the algebraic dual of $B(G)$, where it produced an invariant mean which gave rise to an invariant finitely additive "probability measure" on *all* subsets of G. In Chap. 12, thanks to an enhanced Markov–Kakutani Theorem, the same method produced both invariant means and Haar measure for *solvable* compact groups. By invoking the Invariant Hahn–Banach Theorem (Theorem 12.5, p. 151) we could even produce an invariant mean whose associated finitely additive measure extended Haar measure from the Borel sets to all the subsets of G.

However our luck runs out if we try to extend the Markov–Kakutani Theorem further, in the hope of providing Haar measure for *all* compact groups. In Chap. 11 we saw that not every compact group has an invariant mean; the group $SO(3)$ of rotation matrices, being paradoxical, furnishes just such an example. Thus, at least for $G = SO(3)$, there's no kind spirit to provide an appropriate fixed point for translation adjoints on the algebraic dual of $B(G)$. However the story is different for the *topological* dual of $C(G)$, thanks to the Ryll–Nardzewski Theorem.

To apply that theorem we'll need to know that the continuity previously established for algebraic adjoints remains true of topological ones:

Lemma 13.9. *Suppose X is a Banach space and X^* its topological dual. If T is a continuous linear transformation on X then its topological adjoint T^* is weak-star continuous on X^*.*

Proof. The key here is that T^* is the restriction to X^* of the algebraic adjoint $T^\sharp$ acting on the algebraic dual $X^\sharp$. Proposition 9.18 tells us that $T^\sharp$ is weak-star continuous on $X^\sharp$. Now the weak-star topology on X^* is the restriction of the weak-star topology on $X^\sharp$, so T^* inherits the weak-star continuity of $T^\sharp$.   □

Theorem 13.10. *Haar measure exists for every compact topological group.*

Outline of proof. Let $\mathscr{G}$ denote the group of left-translation operators L_γ for $\gamma \in G$, and let $\mathscr{G}^*$ be the corresponding group of adjoints, operating on $C(G)^*$. Let $\mathscr{K}$ denote the collection of positive linear functionals Λ on $C(G)$ with $\Lambda(1) = 1$.

By Exercise 9.14 (p. 114) we know that $|\Lambda(f)| \leq \|f\|$ for each $\Lambda \in \mathscr{K}$, hence $\mathscr{K}$ is a pointwise bounded subset of $C(G)^*$. By Theorem 9.12 (p. 111), our "infinite dimensional Heine–Borel theorem") $\mathscr{K}$ is therefore relatively compact in the product topology of $\mathbb{R}^{C(G)}$. Now $\mathscr{K}$ is the analogue for $C(G)$ of the set $\mathscr{M}$ of means on $B(G)$, and the proof that $\mathscr{M}$ is closed in $\mathbb{R}^{B(G)}$ works as well to show that $\mathscr{K}$ is closed in $\mathbb{R}^{C(G)}$, hence $\mathscr{K}$ is a compact subset of $\mathbb{R}^{C(G)}$. Since $\mathscr{K}$ is contained in $C(G)^*$, and since the weak-star topology of $C(G)^*$ is just the restriction to that space of the product topology of $\mathbb{R}^{C(G)}$, we see that $\mathscr{K}$ *is weak-star compact in* $C(G)^*$.

Clearly $\mathscr{K}$ is convex. Each of the operators in $\mathscr{G}^*$ is a linear self-map of $\mathscr{K}$ that, by Lemma 13.9, is weak-star continuous on $C(G)^*$. Thus if we can show that $\mathscr{G}^*$ is uniformly injective on $\mathscr{K}$, the Ryll–Nardzewski Theorem will provide a fixed point $\Lambda \in \mathscr{K}$ for $\mathscr{G}^*$. Just as in the commutative and solvable cases, the Riesz Representation Theorem will provide a regular Borel probability measure μ for G that represents Λ via integration, with the $\mathscr{G}^*$-invariance of Λ translating into left G-invariance for μ, i.e., μ will be Haar measure for G.

Proof for G metrizable. By Proposition B.7 we know that $C(G)$ is separable, hence Theorem 13.8, our "lite" version of the Ryll–Nardzewski Theorem, will apply to $C(G)^*$ (taken in its weak-star topology) once we've established that $\mathscr{G}^*$ is uniformly injective. For this we'll need to know that:

(†)   *For each $\Lambda \in \mathscr{K}$ the map $\gamma \to L_\gamma^* \Lambda$ takes G continuously into $\mathscr{K}$ (with its weak-star topology).*

> *Proof.* Let μ denote the regular Borel probability measure for G that—thanks to the Riesz Representation Theorem—represents Λ, i.e., $\Lambda(f) = \int_G f \, d\mu$ for $f \in C(G)$. Since both G and the weak-star topology on $\mathscr{K}$ are metrizable (the latter thanks to Propositions 13.7 and B.7) we may use sequences to establish continuity. Suppose (γ_n) is a sequence in G that converges to an element γ of G. Then for $f \in C(G)$ we have, thanks to the Dominated Convergence Theorem:
>
> $$(L_{\gamma_n}^* \Lambda)(f) = \Lambda(L_{\gamma_n} f) = \int f(\gamma_n x) \, d\mu(x) \to \int f(\gamma x) \, d\mu(x) = (L_\gamma^* \Lambda)(f).$$
>
> Thus $L_{\gamma_n}^* \Lambda \to L_\gamma^* \Lambda$ in the weak-star topology (cf. Exercise 9.10, p.110), which establishes the desired continuity of the map $\gamma \to L_\gamma^*$.   □

To show is that $\mathscr{G}^*$ is uniformly injective on $\mathscr{K}$, fix Φ and Λ in $C(G)^*$ and suppose the zero-functional belongs to the weak-star closure of $\Delta = \{L_\gamma^*\Phi - L_\gamma^*\Lambda : \gamma \in G\}$. Our goal is to prove that $\Phi = \Lambda$. Note that the set Δ is pointwise bounded on $C(G)$, so its closure is weak-star compact, and therefore metrizable. This, along with the metrizability of G allows the following rephrasing of hypothesis on Δ: *There exists a sequence* (γ_n) *of elements of G such that* $L_{\gamma_n}^*\Phi - L_{\gamma_n}^*\Lambda \to 0$ *weak-star in* $C(G)^*$. Since G is compact we may, upon replacing our original sequence of group elements by an appropriate subsequence, assume that (γ_n) converges to an element $\gamma \in G$. By the continuity established in (†) above we have

$$0 = \lim_n[L_{\gamma_n}^*\Phi - L_{\gamma_n}^*\Lambda] = L_\gamma^*\Phi - L_\gamma^*\Lambda = \Phi \circ L_\gamma - \Lambda \circ L_\gamma = (\Phi - \Lambda) \circ L_\gamma.$$

Since L_γ is an isomorphism of $C(G)$ onto itself this implies $\Phi = \Lambda$, as desired.

Thus the hypotheses of our "lite" version of the Ryll–Nardzewski Theorem (Theorem 13.8) are satisfied with $X = C(G)^*$, $K = \mathscr{K}$, and $\mathscr{S} = \mathscr{G}^*$, so $\mathscr{G}^*$ has a fixed point in $\mathscr{K}$; as noted above, this provides Haar measure for G. □

Sketch of proof for arbitrary compact G. In this setting the weak-star topology on $C(G)^*$ is no longer metrizable on every compact set, so we can't use sequential arguments. This means we must modify the continuity proof for $\gamma \to L_\gamma^*$ so as to avoid the Dominated Convergence Theorem. Instead the idea is to first prove that for each $f \in C(G)$ the map $\gamma \to L_\gamma f$ is continuous from G to $C(G)$ in its norm topology. This follows from the fact each continuous function on G exhibits a form of uniform continuity that generalizes the one familiar to us from metric-space theory.[8] Once we've established the desired continuity of the map $\gamma \to L_\gamma f$ it's an easy matter to show that the map $\gamma \to L_\gamma^*\Lambda$ is continuous for each $\Lambda \in C(G)^*$. The rest of the argument then goes through almost word-for-word, with nets and subnets replacing sequences and subsequences.

Now that we know Haar measure exists (uniquely and bi-invariantly) on every compact group, it's time to investigate an important example.

13.5 Haar Measure on SO(3)

SO(3) is the collection of 3×3 matrices whose determinant is 1 and whose columns form an orthonormal subset of $\mathbb{R}^3$. For every real square matrix A, such column orthonormality expresses itself in the matrix equation $AA^t = I$, where A^t denotes the transpose of A, and I is the identity matrix of the size of A. This, along with the multiplicative property of determinants, makes it easy to show that SO(3) is a group under matrix multiplication; in Appendix D it's shown that the elements of this group are precisely the matrices (with respect to the standard basis of $\mathbb{R}^3$) of rota-

---

[8] See, e.g., [103], proof of Theorem 5.13, pp. 129–130.

tions of $\mathbb{R}^3$ about the origin. For topological purposes we'll regard SO(3) as a subset of the sphere of radius $\sqrt{3}$ in $\mathbb{R}^9$. It's easy to check that the Euclidean topology of $\mathbb{R}^9$ makes SO(3) into a compact group (exercise), which therefore possesses Haar measure. What is this measure? How does one integrate respect to it?

What is Haar Measure on SO(3)? Since we can regard SO(3) as the group of rotations of the unit sphere S^2 of $\mathbb{R}^3$, one might suspect that its Haar measure should somehow involve surface area measure on that sphere. A natural way of connecting group with sphere is to define the map $\varphi \colon$ SO(3) $\to S^2$ which takes a matrix $x \in$ SO(3) to its last column. Thus $\varphi(x) = xe_3$, where we regard $\mathbb{R}^3$ as a space of column vectors, and $e_3 = [0, 0, 1]^t$ is the unit vector in $\mathbb{R}^3$ "along the z-axis."

Since matrix entries are continuous functions of their matrices, the map φ is continuous. It is surjective (each unit vector can be the third column of a matrix in SO(3)), but not one-to-one (it's constant on subsets of SO(3) whose elements share the same third column).

More precisely, let K denote the subgroup of matrices in SO(3) that fix the vector e_3 (i.e., which have third column equal to e_3). Then the coset modulo K of a matrix $x \in$ SO(3) is $xK = \{xk : k \in K\}$, namely all matrices in SO(3) with third column the same as that of x. If x and y in SO(3) have different third columns (i.e., belong to different cosets mod K), then $\varphi(x) \neq \varphi(y)$. Thus φ is a one-to-one mapping of cosets mod K onto S^2.[9]

Now suppose $f \in C(\mathrm{SO}(3))$. The subgroup K, being compact, has its own Haar measure which we'll denote by dk. Define f_K on $C(\mathrm{SO}(3))$ by:

$$f_K(x) := \int_K f(xk)dk \qquad (x \in \mathrm{SO}(3)).$$

Clearly f_K is continuous; by the invariance of dk it is constant on cosets of SO(3) modulo K.

To make the definition of f_K more concrete, observe that each element of K has the form

$$k(\theta) = \begin{pmatrix} \cos\theta & -\sin\theta & 0 \\ \sin\theta & \cos\theta & 0 \\ 0 & 0 & 1 \end{pmatrix} \qquad (0 \leq \theta < 2\pi),$$

so the map that takes $k(\theta)$ to its upper left-hand 2×2 submatrix, and then to the unimodular complex number $\cos\theta + i\sin\theta$, establishes a homeomorphic isomorphism between K and the unit circle, now viewed as the group of rotations of $\mathbb{R}^2$. This allows Haar measure on K to be concretely represented by the Haar measure of the circle group, i.e., normalized Lebesgue arc-length measure:

$$\int_K g(k)\,dk = \frac{1}{2\pi} \int_0^{2\pi} g(k(\theta))\,d\theta \qquad (g \in C(K)).$$

---

[9] As such, φ can be regarded as a map taking SO(3) onto the quotient space $\mathrm{SO}(3)/K$, but right now we'll avoid the notion of "quotient space."

Thus for $f \in C(\mathrm{SO}(3))$,

$$f_K(x) = \frac{1}{2\pi} \int_0^{2\pi} f(xk(\theta))\, d\theta \qquad (x \in \mathrm{SO}(3)). \tag{13.3}$$

Exercise 13.4. Define $f \colon \mathrm{SO}(3) \to [-1,1]$ by $f(x) = (x_{2,2})^2$. Show that

$$f_K(x) = \frac{1 - (x_{2,3})^2}{2}$$

for each $x \in \mathrm{SO}(3)$.

The function f_K, being constant on cosets mod K, may be viewed via the map φ as a function on S^2. More precisely, let

$$\hat{f}(p) = f_K\big(\varphi^{-1}(p)\big) \qquad (p \in S^2).$$

Theorem 13.11. *Haar measure dx on $\mathrm{SO}(3)$ is given by*

$$\int_{\mathrm{SO}(3)} f(x)\, dx = \int_{S^2} \hat{f}(p)\, d\sigma(p) \qquad (f \in C(\mathrm{SO}(3))),$$

where σ denotes surface area measure on S^2, normalized to have unit mass.

Proof. Define the linear functional Λ on $C(\mathrm{SO}(3))$ by

$$\Lambda(f) = \int_{S^2} \hat{f}(p)\, d\sigma(p) \qquad (f \in C(\mathrm{SO}(3))).$$

Then Λ is a positive linear functional on $C(\mathrm{SO}(3))$, so the Riesz Representation Theorem provides a regular Borel probability measure μ for $\mathrm{SO}(3)$ such that $\Lambda(f) = \int f\, d\mu$ for all $f \in C(\mathrm{SO}(3))$. One checks easily that if $f \equiv 1$ on $\mathrm{SO}(3)$ then $\Lambda(f) = 1$, i.e., that $\mu(\mathrm{SO}(3)) = 1$; thus μ is a probability measure.

To show that μ is Haar measure on $\mathrm{SO}(3)$ we need only check is that it is left-invariant, i.e., that if L_y is the "left-translation" operator on $C(\mathrm{SO}(3))$:

$$(L_y f)(x) = f(yx) \qquad (f \in C(\mathrm{SO}(3)),\ x,y \in \mathrm{SO}(3)),$$

then $\Lambda(L_y f) = \Lambda(f)$ for $f \in C(\mathrm{SO}(3))$ and $y \in \mathrm{SO}(3)$.

The proof of this hinges on the identity

$$\widehat{(L_y f)} = L_y(\hat{f}) \qquad \text{for each } f \in C(\mathrm{SO}(3)), \tag{13.4}$$

where on the right-hand side we have L_y operating in the obvious way on $C(S^2)$, namely: $L_y g(p) = g(yp)$ for $p \in S^2$ and $g \in C(S^2)$. Granting this: for $f \in C(\mathrm{SO}(3))$ and $y \in \mathrm{SO}(3)$:

$$\Lambda(L_y f) = \int_{S^2} \widehat{L_y f}\, d\sigma = \int_{S^2} L_y(\hat{f})\, d\sigma = \int_{S^2} \hat{f}\, d\sigma = \Lambda(f),$$

where the second equality uses (13.4), and the third one follows from the rotation-invariance of surface area measure on S^2.

The proof of (13.4) involves nothing more than chasing definitions. Fix $p \in S^2$ and choose $x \in SO(3)$ with $\varphi(x) = p$ (so that p is the third column of the matrix x). Fix the "translator" $y \in SO(3)$. Note that

$$\varphi(yx) := yxe_3 = y\varphi(x) = yp \tag{13.5}$$

so for $f \in C(SO(3))$,

$$\widehat{(L_y f)}(p) = (L_y f)_K(\varphi^{-1}(p)) = \int_K (L_y f)(xk)\,dk$$

$$= \int_K f(yxk)\,dk = f_K(yx)$$

$$= f_K(\varphi^{-1}(yp)) \qquad \text{(by (13.5))}$$

$$= \hat{f}(yp) = (L_y \hat{f})(p)$$

which completes the proof of the theorem.   □

It's tempting to think of Theorem 13.11 as somehow expressing Haar measure on $SO(3)$ as the product of normalized surface area on the sphere S^2 and Haar measure on the subgroup K. Not so! One must instead regard Theorem 13.11 as "disintegrating" Haar measure on $SO(3)$ into a family of translates of dk—one for each coset mod K of $SO(3)$—which are "glued together" by the surface area measure $d\sigma$.

More precisely, denote left-multiplication by x on $SO(3)$ by λ_x (i.e., $\lambda_x(y) = xy$ for $y \in SO(3)$), and Haar measure on K by ν. Then for $x \in SO(3)$ can use the change-of-variable formula of measure theory to rewrite the definition of $f_K(x)$ as:

$$f_K(x) = \int_K f(\lambda_x(k))\,d\nu(k) = \int_{xK} f\,d(\nu\lambda_x^{-1}) \qquad (f \in C(SO(3))).$$

We have $xK = \varphi^{-1}(p)$, where p is the third column of the matrix x. Since $f_K(x)$ is constant on $\varphi^{-1}(p)$, the formula above shows that the probability measure $\nu\lambda_x^{-1}$ depends only on the point $p \in S^2$. Upon writing ν_p for this measure we can rewrite the conclusion of Theorem 13.11 as:

$$\int_{SO(3)} f(x)\,dx = \int_{S^2} \left(\int_{\varphi^{-1}(p)} f\,d\nu_p \right) d\sigma(p) \qquad (f \in C(SO(3))), \tag{13.6}$$

which exhibits how Haar measure on $SO(3)$ "disintegrates" into the measures ν_p.

Exercise 13.5. Express the familiar formula from Calculus by which one integrates a continuous real-valued function over the plane triangle $\Delta = \{0 \le y \le x, 0 \le x \le 1\}$ as a similar "disintegration" of Lebesgue area measure on Δ into a family of one dimensional measures on vertical (or, if you wish, the horizontal) cross-sections of that triangle.

Exercise 13.6. Identify Haar measure on O(3), the group of *all* orthogonal 3×3 matrices.

13.6 Computation of some Haar integrals over SO(3)

Let's use the usual subscript notion $x_{i,j}$ to denote the entry in the i-th row and j-th column of a matrix x. Suppose g is a real-valued continuous function on the closed real interval $[-1, 1]$. What is $\int_{SO(3)} g(x_{i,j}) dx$? For $i = j = 3$ the answer is easy to find, since in this case the function $g(x) = g(x_{3,3})$ is already constant on cosets of SO(3) mod K, hence in our characterization of Haar measure on SO(3), $g = g_K$, and therefore

$$\int_{SO(3)} g(x_{3,3}) dx = \int_{S^2} g(p_3) d\sigma(p). \tag{13.7}$$

Let (θ, φ) be the usual spherical coordinates of a point of $p \in S^2$, i.e., φ is the angle from the z-axis to the line from the origin to p, and θ is the angle from the x-axis to that line. Thus $p = [\cos\theta \sin\varphi, \sin\theta \sin\varphi, \cos\varphi]^t$. In particular, $p_3 = \cos\varphi$, and we know from multivariable calculus that (normalized) area measure on S^2 is given by $d\sigma(p) = \frac{1}{4\pi} \sin\varphi \, d\varphi \, d\theta$. Thus the right-hand side of (13.7) is

$$\frac{1}{4\pi} \int_{\theta=0}^{2\pi} \left(\int_{\varphi=0}^{\pi} g(\cos\varphi) \sin\varphi \, d\varphi \right) d\theta,$$

so upon setting $t = -\cos\varphi$ in the inner integral we obtain

$$\int_{SO(3)} g(x_{3,3}) dx = \frac{1}{2} \int_{-1}^{1} g(t) dt. \tag{13.8}$$

The same reasoning could be used to integrate $g(x_{i,3})$ for $i = 1, 2$, but there's no reason to do so; the bi-invariance of Haar measure reduces all such integrals to the one we just worked out, yielding

Proposition 13.12. *Suppose* $g \in C([-1, 1])$ *and* $1 \le i, j \le 3$. *Then*

$$\int_{SO(3)} g(x_{i,j}) dx = \frac{1}{2} \int_{-1}^{1} g(t) dt.$$

Proof. One can find matrices $a, b \in SO(3)$ such that $x_{i,j} = (axb)_{3,3}$ (exercise) whereupon the bi-invariance of Haar measure on SO(3) yields

$$\int_{SO(3)} g(x_{i,j}) dx = \int_{SO(3)} g((axb)_{3,3}) dx = \int_{SO(3)} g(x_{3,3}) dx.$$

This, along with (13.8), above gives the promised result. $\qquad\square$

Corollary 13.13. $\int_{SO(3)} x_{i,j} dx = 0$ *for all* (i, j) *with* $1 \le i, j \le 3$.

Exercise 13.7. Show that the "normalized" matrix entries $\{x_{i,j}/\sqrt{3} : 1 \leq i, j \leq 3\}$ form an orthonormal set in $L^2(\mathrm{SO}(3))$ (with respect to Haar measure).

Suggestion. To prove orthogonality, use the bi-invariance of Haar measure to reduce the problem to showing that $x_{i,j} \perp x_{3,3}$ whenever $(i,j) \neq (3,3)$. For this you'll need to show that if $f(x) = x_{i,j}$ then $f_K(x) = 0$ whenever $j \neq 3$, and $= x_{i,3}$ otherwise.

Characters again. Recall from Sect. 9.7 the notion of *character* for a topological group: a continuous homomorphism of that group into the circle group $\mathbb{T}$. We saw in that section that characters form the basis of an extension to compact *abelian* groups of Fourier analysis on the circle. The exercise below shows that the situation is much different for non-commutative groups.

Exercise 13.8 (The character group of SO(3) is trivial). This exercise requires only the fact that matrices in SO(3) are in one-to-one correspondence with rotations of $\mathbb{R}^3$ about the origin, that rotations preserve lengths of vectors and angles between vectors, and that each such rotation is uniquely determined by its axis (a line through the origin, each point of which is fixed) and its angle of rotation about that axis. For full details see Appendix D.

Let $R_u(\theta)$ denote the rotation having axis in the direction of the unit vector u and rotation angle $\theta \in [-\pi, \pi)$, where the sign of the angle is determined by the "right-hand rule." If u is the unit vector along the x-axis, we'll write $R_x(\theta)$ instead of $R_u(\theta)$, and similarly for $R_z(\theta)$.

(a)   Suppose u, v is a pair of unit vectors in $\mathbb{R}^3$, and $M \in \mathrm{SO}(3)$ maps u on to v. Then for each angle θ we have the (unitary) similarity $R_v(\theta) = MR_u(\theta)M^{-1}$ (see Appendix D, p. 205 for a more detailed version of this). Conclude that for each character γ on SO(3), the value at each matrix in SO(3) depends only on the angle of rotation and not on the axis.

(b)   Let $M(\theta) = R_z(\theta)R_x(-\theta)$. Show that if γ is a character of SO(3) then $\gamma(M(\theta)) = 1$ for every $\theta \in [-\pi, \pi)$. Thus one need only prove that every $\alpha \in [-\pi, \pi)$ is the angle of rotation of some $M(\theta)$ or its inverse.

(c)   Prove that $\cos \theta = \big(\mathrm{trace}\,(R_u(\theta)) - 1\big)/2$. Use this to show that if $f(\theta)$ is the cosine of the angle of rotation of $M(\theta)$ then

$$f(\theta) = -1/4 + \cos(t) + \cos(2t)/4.$$

Show that f maps the interval $(-\pi, \pi]$ onto $[-1, 1]$. Conclude that if $\beta \in (-\pi, \pi]$ then there exists $\theta \in (-\pi, \pi]$ such that either $M(\theta)$ or its inverse is a rotation through angle β. Thus SO(3) has only the trivial character $\gamma \equiv 1$.

13.7 Kakutani's Equicontinuity Theorem

The Ryll–Nardzewski Theorem generalizes:

Corollary 13.14 (Kakutani). *Suppose K is a nonvoid, compact, convex subset of a normed linear space, and $\mathscr{G}$ is an equicontinuous group of affine self-maps of K. Then $\mathscr{G}$ has a fixed point in K.*

Proof. By Ryll–Nardzewski's Theorem it's enough to prove that the group $\mathscr{G}$ is uniformly injective. To this end suppose x and y are vectors in K with $x \neq y$. Let $\varepsilon = \|x - y\|$. By equicontinuity there exists $\delta > 0$ such that if v and w are vectors in K with $\|v - w\| < \delta$ then $\|A(v) - A(w)\| < \varepsilon$ for every $A \in \mathscr{G}$. Now fix $A \in \mathscr{G}$, so A^{-1} also belongs to $\mathscr{G}$ and $\varepsilon = \|A^{-1}A(x) - A^{-1}A(y)\|$, hence $\|A(x) - A(y)\|$ must be $\geq \delta$. Thus the zero vector does not belong to the closure of $\{A(x) - A(y) : A \in \mathscr{G}\}$, which establishes the uniform injectivity of $\mathscr{G}$. □

This proof, with the notion of "equicontinuity" suitably interpreted, can be made to work as well in every locally convex topological vector space; see, for example, [103, Theorem 5.11, pp. 127–128]. Kakutani's theorem can be used as the first step of a proof (much different from the one given above) of the existence of Haar measure for every compact group. See, e.g., [103, Theorems 5.13–5.14, pp. 129–132].

Notes

The "real" Ryll–Nardzewski Theorem. Our version of Ryll–Nardzewski's theorem (Theorem 13.1) is due to Hahn [45]; it's a special case of what Ryll–Nardzewski actually proved. The "real" result, proved in [105], assumes compactness of the convex set K and continuity for the affine semigroup of maps $\mathscr{S}$ for the *weak topology* induced on the locally convex space X by its dual space. The notion of "uniform injectivity" however still refers to the original topology of X. Shorter proofs were subsequently given by Namioka and Asplund [82], and later by Dugundji and Granas [35].

Nets. A *sequence* $(x_n)_1^{\infty}$ from a set X is just a function x from the set $\mathbb{N}$ of natural numbers to X, with x_n denoting the value $x(n)$. More generally, suppose D is a set on which there is a relation $\prec$ that is both reflexive and transitive, and for which every pair of elements in A has an upper bound. The pair $(A, \prec)$ is called a *directed set*, and a function $x \colon A \to X$ is called a *net* from X, often abbreviated $(x_\delta)_{\delta \in D}$. If X is a topological space then to say such a net *converges* to an element $x_0 \in X$ means that for every neighborhood U of x_0 there exists $\delta_0 \in D$ such that $\delta_0 \prec \delta \implies x_\delta \in U$. With this definition, the sequential arguments that establish the properties of closure and continuity for metric spaces can be carried over directly to general topological spaces simply by replacing sequences with nets. The subsequential characterization of compactness for metric spaces even has an analogue for nets in general topological spaces, but for this to happen the proper definition of "subnet" must be a lot more subtle than that of "subsequence." For the details see, e.g., [123, Chap. 4].

Haar Measure is named for the Hungarian mathematician Alfred Haar (1885–1933) whose landmark paper [42, 1933] proved its existence for metrizable locally compact groups. Subsequently Banach [8, 1937] modified Haar's argument to provide measures invariant for the action of compact transformation groups acting continuously on compact metric spaces. The existence of an invariant measure for each locally compact (Hausdorff) topological group was proved in 1940 by André Weil. Predating all of this, in 1897 Adolph Hurwitz defined the notion of invariant integral for SO(n), essentially identifying Haar measure for that group. For this, and further historical background and references, see Hawkins' exposition [49, 1999] (especially p. 185 for Hurwitz's result, and pp. 194–196 for the rest). Diestel and Spalsbury in [29, 2014] provide a recent and accessible account of Haar measure, its history, and many of its applications including a nice introduction to its role in harmonic analysis on compact groups as well as some recent applications to Banach space theory.

Disintegration of Haar Measure with respect to a subgroup. The argument that proved Theorem 13.11 goes through almost *verbatim* to prove the same result with SO(3) replaced by a compact group G, K by a closed subgroup, S^2 by G/K, and σ by $\mu\pi^{-1}$, where $\pi\colon G \to G/K$ is the "quotient map" $x \to xK$ $(x \in G)$. The point is that the quotient space G/K has a natural topology that renders it compact and Hausdorff (namely: the strongest topology that makes π continuous); in the case of SO(3) this is just the topology induced on G/K by its identification with S^2. One also needs to note that the natural action of G on G/K $(xK \to gxK$ for $g, x \in K)$ is continuous in this topology. With these substitutions Theorem 13.11 remains true, and signals a disintegration of μ with respect to the family of translates v_p of v to the cosets p that make up G/K.

Exercise 13.8. The argument outlined for this exercise expands on that of [36, Sect. 4.8.4, p. 232].

Appendices

These appendices are intended to provide a convenient reference for some pre-requisite background material. Use only as needed!

Appendix A
Advanced Calculus

A.1 Differentiation in $\mathbb{R}^N$

Here we'll review those parts of the theory of differentiation of vector-valued functions of several variables needed for our proof of the Brouwer Fixed-Point Theorem. For more details see, e.g., [101, Chap. 9], [2, Chap. 6] or the freely available online textbooks [110] and [118].

Definition and Basic Properties. We'll think of $\mathbb{R}^N$ as a space of column vectors, with $\mathbb{R}^N$-valued functions on a subset V of $\mathbb{R}^N$ to be thought of as column vectors with each component a real-valued function on V. If V is an open subset of $\mathbb{R}^N$, to say that a function $f: V \to \mathbb{R}^N$ is *differentiable* at a point $x_0 \in V$ means that there is a linear transformation on $\mathbb{R}^N$ (which we denote by $f'(x_0)$) such that

$$\lim_{h \to 0} \frac{|f(x_0 + h) - f(x_0) - f'(x_0)h|}{|h|} = 0. \tag{A.1}$$

Suppose f is differentiable at x_0. Then it's clear from the definition that f is continuous at x_0. Furthermore, upon letting $h = te_j$ where e_j is the j-th standard basis vector for $\mathbb{R}^N$ (the vector with 1 in the j-th position and zeros elsewhere) and t is real, we have from the definition above:

$$\lim_{t \to 0} \frac{f(x_0 + te_j) - f(x_0)}{t} = f'(x_0)e_j.$$

Thus $f'(x_0)e_j$, the j-th column of our matrix, is the partial derivative $(\partial f / \partial x_j)(x_0)$ of the vector-valued function f with respect to the j-th variable. Consequently the derivative $f'(x_0)$ is uniquely defined by (A.1), each coordinate function f_i is differentiable at x_0, and with respect to the standard basis of $\mathbb{R}^N$ the matrix of $f'(x_0)$ has:

- As its j-th column the partial derivative of f with respect to its j-th variable,
- As its i-th row the gradient of the coordinate function f_i, and
- As its (i, j)-th element the partial derivative $(\partial f_i / \partial x_j)(x_0)$.

© Springer International Publishing Switzerland 2016
J.H. Shapiro, *A Fixed-Point Farrago*, Universitext,
DOI 10.1007/978-3-319-27978-7

There is a partial converse[1]:

> Suppose f is an $\mathbb{R}^N$-valued function defined on a neighborhood V of a point $x_0 \in \mathbb{R}^N$, and that each partial derivative $(\partial f_i / \partial x_j)(x_0)$ exists $(i, j = 1, 2, \ldots, N)$ *and is continuous on* V. Then f is differentiable at x_0, and f' is continuous on V.

Here the continuity of f' at a point $x \in V$ can be interpreted in several equivalent ways. Perhaps easiest is to demand for every vector $h \in \mathbb{R}^N$ that $f'(x_k)h \to f'(x)h$ in $\mathbb{R}^N$ whenever $x_k \to x$ in V. Equivalently, we may, for $x \in V$, identify $f'(x)$ with its matrix with respect to the standard basis and demand that each matrix entry be continuous on V, or equivalently that f', viewed as a mapping $V \to \mathbb{R}^{N^2}$, be continuous.

The Chain Rule and Some Consequences

Theorem A.1 (The Chain Rule). *Suppose f and g are $\mathbb{R}^N$-valued functions, with f defined on a neighborhood of $x_0 \in \mathbb{R}^N$ and g defined on a neighborhood of $f(x_0)$. If f is differentiable at x_0 and g is differentiable at $f(x_0)$ then $g \circ f$ is differentiable at x_0 and $(g \circ f)'(x_0) = g'(f(x_0))f'(x_0)$, where on the right-hand side we see a composition of linear transformations.*

The proof is almost identical with that of the one-variable case. For the details see [101, Theorem 9.15, p. 214].

Theorem A.2 (The Mean-Value Inequality). *Suppose f is an $\mathbb{R}^N$-valued function defined on an open subset V of $\mathbb{R}^N$, that $f \in C^1(V)$, and that K is a compact, convex subset of V. Then there exists a positive constant M such that*

$$|f(y) - f(x)| \leq M|y - x|$$

for every pair x, y of points of K.

Proof. Let $M = \max_{x \in K} \|f'(x)\|$, where the norm of $f'(x)$ is defined to be

$$\|f'(x)\| = \left[\sum_{i,j} \left(\frac{\partial f_i}{\partial x_j}(x) \right)^2 \right]^{\frac{1}{2}},$$

the norm that results when the matrix of $f'(x)$ is viewed as a vector in $\mathbb{R}^{N^2}$.

Define $\gamma: [0, 1] \to \mathbb{R}^N$ by $\gamma(t) = (1 - t)x + ty$ for $0 \leq t \leq 1$. Thus $\gamma(0) = x, \gamma(1) = y$, and $\gamma' \equiv y - x$. The convexity of K insures that $\gamma([0, 1]) \subset K$, so $g = f \circ \gamma$ maps $[0, 1]$ into K, hence

---

[1] See, e.g., [101, Theorem 9.21, p. 219].

$$|f(y) - f(x)| = |g(1) - g(0)| = \left| \int_0^1 g'(t)\, dt \right|$$

$$\leq \int_0^1 |g'(t)|\, dt = \int_0^1 |f'(\gamma(t))\, \gamma'(t)|\, dt$$

$$\leq \int_0^1 |f'(\gamma(t))|\, |\gamma'(t)|\, dt \leq M|y - x| \qquad \qquad \square$$

Some explanation is needed in the calculation above. In the first line the functions being integrated are $\mathbb{R}^N$-valued; the integrals are vectors obtained by integrating each coordinate of the integrand. In the second line the inequality obtained by passing the norm through the integral sign is Exercise A.1 below, while the following equality comes from the Chain Rule. The first inequality in the third line follows from our definition of matrix norm and the Cauchy–Schwarz inequality, while the final inequality comes from the definition of the constant M and the fact that $\gamma' \equiv y - x$.

Exercise A.1. Suppose $h\colon [0,1] \to \mathbb{R}^N$ is continuous. Show that $\left| \int_0^1 h(t)\, dt \right| \leq \int_0^1 |h(t)|\, dt$.

Suggestion: For each vector $x \in \mathbb{R}^N \setminus \{0\}$ we have (trivially) that $|x| = \langle x, w \rangle$ where $w = \frac{x}{|x|}$.
Use this with $x = \int_0^1 h(t)\, dt$, which can be assumed to be non-zero.

Theorem A.3 (The Inverse-Function Theorem). *Suppose V is an open subset of $\mathbb{R}^N$, $x_0 \in V$, and $f\colon V \to \mathbb{R}^N$ is a C^1-map for which the derivative $f'(x_0)$ is invertible (i.e., for which $\det f'(x_0) \neq 0$). Then there is a neighborhood of x_0 upon which the restriction of f is a homeomorphism with C^1 inverse.*

For the proof, see, e.g., [101, Theorem 9.24, p. 221].

A.2 Approximation by Smooth Functions

In Sect. 4.3 our proof of the Brouwer Fixed-Point Theorem required that continuous, real-valued functions on the unit ball B of $\mathbb{R}^N$ be uniformly approximated by functions having continuous derivatives. Here is a proof of this fact.

Suppose $f\colon B \to \mathbb{R}$ is continuous. By Exercise A.2 below it's enough to assume that f extends to a function continuous on all of $\mathbb{R}^N$, with compact support.

The rest of the proof begins with a C^1 "bump function." Let φ be a non-negative C^1 function on $\mathbb{R}^N$ supported in B, with $\int \varphi = 1$ (here unadorned integrals extend over all of $\mathbb{R}^N$). For example, take $\varphi(x)$ to be $\frac{2}{\text{vol}(B)} \cos^2(\frac{\pi}{2}|x|)$ when $|x| \leq 1$, and 0 when $|x| > 1$.

Now fix $\delta > 0$ (to be later specified precisely) and set $\varphi_\delta(x) = \delta^{-N} \varphi(x/\delta)$. Then φ_δ has the same properties as φ (C^1, non-negative, compact support, integral = 1), but now its support lies in $\delta B = \{x \in \mathbb{R}^N : |x| \leq \delta\}$. Define

$$g(x) = \int f(t)\varphi_\delta(x - t)\, dt \qquad (x \in \mathbb{R}^N), \tag{A.2}$$

where the integral on the right (and all further integrals in this proof) are understood to extend over all of $\mathbb{R}^N$. "Differentiation under the integral sign" (cf. [101, Theorem 9.42, pp. 236–237]) shows that g is a C^1 function on $\mathbb{R}^N$. Since $\int \varphi_\delta = 1$ the same is true of the t-integral of $\varphi_\delta(x-t)$, so we have for $x \in K$:

$$|f(x) - g(x)| = \left| \int [f(x) - f(t)] \varphi_\delta(x-t)\,dt \right| \leq \int |f(x) - f(t)| \varphi_\delta(x-t)\,dt \quad \text{(A.3)}$$

In the estimate above the integrands are supported in the ball

$$B_{x,\delta} := x + \delta B = \{t \in \mathbb{R}^N : |x-t| < \delta\}$$

so the integrals extend only over that ball, hence

$$\int |f(x) - f(t)| \varphi_\delta(x-t)\,dt \leq \omega(f,\delta) := \max\{|f(x) - f(t)| : x,t \in \mathbb{R}^N\}. \quad \text{(A.4)}$$

Because f has compact support it is uniformly continuous on $\mathbb{R}^N$, so $\omega(f,\delta)$ is finite for each $\delta > 0$ and $\to 0$ as $\delta \to 0$. Thus for each $x \in \mathbb{R}^N$ we obtain from the estimates (A.3) and (A.4) above:

$$|f(x) - g(x)| \leq \omega(f,\delta) \int \varphi_\delta(x-t)\,dt = \omega(f,\delta)$$

which, upon choosing δ so that $\omega(f,\delta) < \varepsilon$, insures that g provides the desired uniform approximation to f, even on all of $\mathbb{R}^N$. □

Exercise A.2. Show that: given $f : B \to \mathbb{R}$ continuous and $\varepsilon > 0$, there exists a continuous, compactly supported function f_0 on $\mathbb{R}^N$ such that $|f(x) - f_0(x)| < \varepsilon$ for every $x \in B$.

Suggestion: First show that for $r > 1$, but sufficiently close to 1, the function $f_1 : x \to f(x/r)$ approximates f to within ε and is continuous on the ball rB. Next, create a function ψ, continuous on $\mathbb{R}^N$ and supported on rB such that $\psi \equiv 1$ on B and $\psi \equiv 0$ off rB. Let $f_0 = \psi f_1$.

A.3 Change-of-Variables in Integrals

Here is the fundamental result about changing variables in Riemann integration of functions of several variables.

Theorem A.4 (The Change-of-Variable Theorem). *Suppose φ is an $\mathbb{R}^N$-valued C^1 mapping of an open subset of $\mathbb{R}^N$ that contains a compact, connected subset K whose boundary has volume zero. Suppose further that on K the mapping φ is one-to-one and that $\det \varphi'$ is never zero. Then for every continuous, real-valued function f defined on $\varphi(K)$:*

$$\int_{\varphi(K)} f(y)\,dy = \int_K f(\varphi(x)) |\det \varphi'(x)|\,dx$$

The "volume-zero" condition in the hypotheses means that for every $\varepsilon > 0$ the boundary of K can be covered by open "boxes" with sides parallel to the coordinate axes (i.e., N-fold cartesian products of intervals), the sum of whose volumes is $< \varepsilon$. This is precisely the condition needed to insure that every real-valued continuous function on K is Riemann integrable.

Notes

Approximation by smooth functions. The integral defining the smooth approximation g in (A.2) is called the *convolution* of f and φ_δ, written $g = f * \varphi_\delta$. Exercise 9.5 also featured a convolution integral in the context of topological groups.

Regarding Exercise A.2. Thanks to the Tietze Extension Theorem (see, e.g., [102, Sect. 20.4, p. 389]) the extension promised by this exercise exists with B replaced by any compact subset of $\mathbb{R}^N$.

The change-of-variable formula. A proof of the theorem as stated above can be found in many places, e.g., Apostol's classic text [2, Theorem 10.30, p. 271], or the textbooks of Shurman [110, Theorem 6.7.1, p. 313], and Trench [118, Theorem 7.3.8, p. 496], which are freely available online.

Some authors finesse the hypothesis of zero boundary-volume by demanding that the function f have compact support and that the integrals on both sides of the formula extend over all of $\mathbb{R}^N$; see, e.g., [101, Theorem 10.9, p. 252] for this point of view.

Appendix B
Compact Metric Spaces

We introduced the definition of metric space in Sect. 3.1, p. 27. In what follows we'll be working in a metric space (X,d), which we'll usually just call "X".

Notation. For a point $x_0 \in X$, and a positive real number r, let $B(x_0,r)$ denote the open ball of radius r in X centered at x_0, i.e., $B(x_0,r) = \{x \in X : d(x,x_0) < r\}$.

B.1 ε-Nets and Total Boundedness

Definition B.1 (ε-net). For $\varepsilon > 0$ and $E \subset X$, an *ε-net* is a finite subset F of X with the property that $E \subset \bigcup_{c \in F} B(c,\varepsilon)$.

In other words: An ε-net is a finite subset $F \subset X$ that is "ε-dense" in E in the sense that each point of E lies within ε of some point of F.

Definition B.2. To say a subset E of a metric space is *relatively compact* means that its closure is compact.

Proposition B.3. *If a subset of a metric space is relatively compact then it has, for every $\varepsilon > 0$, an ε-net.*

Proof. Suppose E is a relatively compact subset of X, so that $\overline{E}$, the closure of E in X, is compact. Let $\varepsilon > 0$ be given. Each point of $\overline{E}$ lies within ε of some point of E, i.e., the collection of balls $\{B(e,\varepsilon) : e \in E\}$ is an open cover of $\overline{E}$, so by compactness there is a finite subcover. The centers of the balls in this subcover form the desired ε-net for E. □

Definition B.4. To say a metric space is *totally bounded* means that it has an ε-net for every $\varepsilon > 0$.

With this definition Proposition B.3 can be rephrased:

> *If a subset of a metric space is relatively compact then it is totally bounded.*

© Springer International Publishing Switzerland 2016
J.H. Shapiro, *A Fixed-Point Farrago*, Universitext,
DOI 10.1007/978-3-319-27978-7

The converse of Corollary B.3 is not true in general. *Example*: the set of rationals in the closed unit interval is closed in itself, not compact, but still totally bounded. However, as the Exercise below shows, the converse *does* hold for *complete* metric spaces.

> **Exercise B.1.** In a *complete* metric space, if a subset is totally bounded, then it is relatively compact.
>
> **Suggestion:** Suppose our subset S is totally bounded, so that for each positive integer n there is a $1/n$-net S_n in S. Fix a sequence of elements in S; we wish to find a subsequence that is convergent in the ambient metric space X. Show by a diagonal argument that there is a subsequence of the original which, for each n, lies eventually within $1/n$ of some element of S_n. Show that this subsequence is Cauchy, hence by completeness, convergent.

B.2 Continuous Functions on Compact Spaces

Definition B.5 (Partition of unity). For an n-tuple $\mathscr{U} = (U_1, U_2, \ldots U_n)$ of open sets that cover a metric space X, a *partition of unity subordinate to* $\mathscr{U}$ is an n-tuple $(p_1, p_2, \ldots p_n)$ of continuous functions $X \to [0,1]$, the j-th one vanishing off U_j, for which the totality sums to 1 on X.

Proposition B.6. *Every finite open cover of a compact metric space has a subordinate partition of unity.*

Proof. As usual, denote our metric space by X, and its metric by d. Note that $d : X \times X \to [0, \infty)$ is continuous, and—because of the compactness of X—bounded (this is our only use of compactness here). Suppose $\mathscr{U} = (U_1, U_2, \ldots U_n)$ is our finite open cover of X. Define $d_j : X \to [0, \infty]$ by

$$d_j(x) = \operatorname{dist}(x, X \setminus U_j) := \inf_{\xi \notin U_j} d(x, \xi) \qquad (x \in X)$$

The boundedness and continuity of d insures that d_j is a continuous function $X \to [0, \infty)$, and d_j vanishes on U_j (note that if we're working with intervals of the real line, then the graph of d_j is a "tent" over U_j). Thus the collection of functions

$$p_j := \frac{d_j}{\sum_{k=1}^n d_k} \qquad (j = 1, 2, \ldots n)$$

is easily seen to have the desired properties (the denominator above never vanishes because $d_k > 0$ on V_k, and the V_k's cover X). $\qquad \square$

Proposition B.7 (Separability). *If X is a compact metric space then both X and $C(X)$ are separable.*

Proof. To see that X is separable, for each positive integer n cover X by the collection of all open balls of radius $1/n$; by compactness this open cover has a finite subcover $\mathscr{B}_n$. Let S_n' be the collection of centers of the balls in $\mathscr{B}_n$; this is a finite set with the property that each point of X lies within $1/n$ of one of its points. Thus $\cup_n S_n$ is a countable dense subset of X.

As for the separability of $C(X)$, we know from Proposition B.6 that there is a partition of unity $\mathscr{P}_n = \{p_1, p_2, \ldots, p_N\}$ on X subordinate to the covering $\mathscr{B}_n$. The countable dense subset of $C(X)$ we seek is going to be the collection of rational linear combinations of vectors in $\bigcup_n \mathscr{P}_n$. To see why this is true, fix $f \in C(X)$ and let $\varepsilon > 0$ be given. Since f is continuous on the compact metric space G it is *uniformly continuous* so there exists $\delta > 0$ such that if $x, y \in X$ with $d(x,y) < \delta$ then $|f(x) - f(y)| < \varepsilon$. Choose a positive integer $n > 1/\delta$ and let x_j be the center of the ball $U_j \in \mathscr{B}_n$. Define $g \in C(G)$ by $g = \sum_{j=1}^{N} f(x_j) p_j$. Then for $x \in G$

$$|f(x) - g(x)| = \left| \sum_{j=1}^{N} [f(x) - f(x_j)] p_j(x) \right| \leq \sum_{j=1}^{N} \underbrace{|f(x) - f(x_j)|}_{<\varepsilon \text{ on } U_j} \underbrace{p_j(x)}_{\equiv 0 \text{ off } U_j} < \varepsilon,$$

i.e., $\|f - g\| < \varepsilon$. To finish the proof, go back to the linear combination defining g and replace each coefficient $f(x_j)$ by a rational number sufficiently close that the new g still lies within ε of f in $C(X)$. $\qquad\square$

Equicontinuity. Suppose X is a metric space with metric d. To say a subset E of $C(X)$ is *equicontinuous* means that: for every $\varepsilon > 0$ there exists $\delta > 0$ (which depends *only* on ε) such that for $x, y \in X$:

$$d(x,y) < \delta \implies |f(x) - f(y)| < \varepsilon \quad \forall f \in E.$$

Theorem B.8 (Arzela–Ascoli, 1883–1885). *If X is a compact metric space then every bounded, equicontinuous subset of $C(X)$ is relatively compact.*

Proof. Let X be a compact metric space and B a bounded, equicontinuous subset of $C(X)$. Suppose $(f_n)_1^{\infty}$ is a sequence in B. We desire to show that there is a subsequence that converges in $C(X)$, i.e., uniformly on X.

To this end let S be a countable dense subset of X (which we know exists by Proposition B.7) and enumerate its elements as $(s_1, s_2, \ldots)$. The first order of business is to find a subsequence of (f_n) that converges pointwise on S. This follows from a standard diagonal argument. By the boundedness of B in $C(I)$ the numerical sequence $(f_n(s_1))_{n=1}^{\infty}$ is bounded, so by Bolzano-Weierstrass it has a convergent subsequence, which we'll write using double subscripts: $(f_{1,n}(s_1))_{n=1}^{\infty}$. Now the numerical sequence $(f_{1,n}(s_2))_{n=1}^{\infty}$ is bounded, so it has a convergent subsequence $(f_{2,n}(s_2))_{n=1}^{\infty}$. Note that the sequence of functions $(f_{2,n})_{n=1}^{\infty}$, since it is a subsequence of $(f_{1,n})_{n=1}^{\infty}$, converges at both s_1 and s_2. Proceeding in this fashion we obtain a countable collection of subsequences of our original sequence:

$$f_{1,1} \; f_{1,2} \; f_{1,3} \; \cdots$$

$$f_{2,1} \; f_{2,2} \; f_{2,3} \; \cdots$$

$$f_{3,1} \; f_{3,2} \; f_{3,3} \; \cdots$$

$$\vdots \quad \vdots \quad \vdots \quad \ddots$$

where the sequence in the nth row converges at the points $s_1, \ldots, s_n$, and each row is a subsequence of the one above it. Thus the *diagonal sequence* $(f_{n,n})$ is a subsequence of the original sequence (f_n), and it converges at each point of S.

For simplicity of notation let $g_n = f_{n,n}$. Let $\varepsilon > 0$ be given, and choose $\delta > 0$ by the equicontinuity of the set B to which these functions belong. Thus $d(x, y) < \delta$ implies $|g_n(x) - g_n(y)| < \varepsilon/3$ for each $x, y \in X$ and each positive integer n. Since the sequence (g_n) converges at every point of S there exists for each $s \in S$ a positive integer $N(s)$ such that

$$m, n > N(s) \implies |g_n(s) - g_m(s)| < \varepsilon/3. \tag{*}$$

The open balls in X with centers in S and radius δ cover X, so there is a finite subcover with centers in a finite subset S_δ of S. Let $N = \max\{N(s): s \in S_\delta\}$. Fix $x \in X$. Then x lies within δ of some $s \in S_\delta$, so if $n, m > N$:

$$|g_n(x) - g_m(x)| \le |g_n(x) - g_n(s)| + |g_n(s) - g_m(s)| + |g_m(s) - g_m(x)|.$$

The first and last terms on the right are $< \varepsilon/3$ by our choice of δ (which was possible because of the equicontinuity of the original sequence), and the same estimate holds for the middle term by our choice of N in (*). In summary: given $\varepsilon > 0$ we have produced N so that for each $x \in X$,

$$m, n > N \implies |g_n(x) - g_m(x)| < \varepsilon/3 + \varepsilon/3 + \varepsilon/3 = \varepsilon.$$

Thus the subsequence (g_n) of (f_n) is Cauchy in $C(X)$, hence convergent there.   □

Notes

Arzela–Ascoli converse. If X is a compact metric space and B is a relatively compact subset of $C(X)$ then B is bounded and equicontinuous (exercise).

Appendix C
Convex Sets and Normed Spaces

C.1 Convex Sets

Suppose V is a real vector space. In Sect. 1.6 we defined a subset C of V to be *convex* provided that whenever x and y are points of C then so is $tx + (1-t)y$ for each real number t with $0 \le t \le 1$ (Definition 1.4, p. 10). The empty set is trivially convex, as is every singleton. It's an easy exercise to check that the intersection of a family of convex sets is convex.

Definition C.1. A *convex combination* of vectors $x_1, x_2, \dots, x_N$ in V is a linear combination $\sum_{j=1}^{N} \lambda_j x_j$ where $(\lambda_1, \lambda_2, \dots, \lambda_N)$ is an N-tuple of non-negative real numbers that sum to 1 (i.e., a vector the *standard simplex* Π_N of $\mathbb{R}^N$, introduced in Sect. 1.7).

Proposition C.2. *A subset of V is convex if and only if it contains every convex combination of its vectors.*

Proof. Since for $0 \le t \le 1$ the sum $tx + (1-t)y$ is a convex combination of the vectors x and y, a set that contains *every* convex combination of its vectors is surely convex. Conversely, suppose C is a convex subset of V and $x_1, x_2, \dots, x_N \in C$. Consider a convex combination $x = \sum_{j=1}^{N} \lambda_j x_j$ of these vectors. *To show:* $x \in C$.

  We'll prove this by induction on N. It's trivial for $N = 1$, and the case $N = 2$ follows from the definition of convexity. So suppose $N > 2$ and we know that every convex combination of $N-1$ vectors in C also belongs to C. Let $t = \sum_{j=1}^{N-1} \lambda_j$ so that $0 \le t \le 1$ and $\lambda_N = 1 - t$. Suppose $t \ne 0$ (else $x = x_N$, so trivially $x \in C$). Then $x = ty + (1-t)x_N$ where $y = \sum_{j=1}^{N-1} (\lambda_j/t)x_j$ is a convex combination of $N-1$ vectors in C, hence belongs to C by our induction hypothesis. Thus $x \in C$. $\qquad\square$

Definition C.3 (Convex Hull). If S is a subset of V then the *convex hull* of S, denoted $\operatorname{conv} S$, is the intersection of all the convex sets that contains S.

  The collection of such convex sets contains V itself, so is nonempty, and we've noted above that the intersection of convex sets is convex. Thus $\operatorname{conv} S$ is the smallest

© Springer International Publishing Switzerland 2016
J.H. Shapiro, *A Fixed-Point Farrago*, Universitext,
DOI 10.1007/978-3-319-27978-7

convex subset of V that contains S. With this definition, Proposition C.2 can be rephrased: *A subset C of V is convex if and only if $C = \operatorname{conv} C$.*

Proposition C.4. *The convex hull of a subset S of V is the collection of all convex combinations of vectors in S.*

Proof. Let $\hat{S}$ denote the collection of all convex combinations of vectors in S. Clearly $\hat{S}$ contains S. To see why $\hat{S}$ is convex, suppose x and y are vectors therein. We may assume each is a convex combination of vectors $x_1, x_2, \ldots, x_N$ in S, say $x = \sum_{j=1}^{N} \lambda_j x_j$ and $y = \sum_{j=1}^{N} \mu_j x_j$. Suppose $0 < t < 1$. Then $tx + (1-t)y = \sum_{j=1}^{N} \eta_j x_j$ where for each index j we have $\eta_j = t\lambda_j + (1-t)\mu_j$. Thus $(\eta_1, \eta_2, \ldots, \eta_n) \in \Pi_N$ so $tx + (1-t)y \in \hat{S}$, proving the convexity of $\hat{S}$. To prove that $\hat{S}$ is the *smallest* convex set containing S, suppose C is convex and $C \supset S$. Then by Proposition C.2 we know that C contains all the convex combinations of its vectors, hence in particular all the vectors in $\operatorname{conv} S$. □

Example. The standard simplex Π_N (see Definition 1.7, page 11) is the convex hull of the standard unit vectors $e_1, \ldots, e_N$ in $\mathbb{R}^N$.

C.2 Normed Linear Spaces

Norms. A normed linear space is a real or complex vector space X upon which is defined a function $\| \cdot \| : X \to [0, \infty)$ that is

(a)   Subadditive: $\|x + y\| \leq \|x\| + \|y\|$ for all $x, y \in X$,
(b)   Positively homogeneous: $\|tx\| = |t| \|x\|$ for every $x \in X$ and scalar t, and for which
(c) $\|x\| = 0$ if and only if $x = 0$.

Examples of norms are:

- The *Euclidean norm* on $\mathbb{R}^N$ or $\mathbb{C}^N$.
- More generally, the norm induced on any real or complex vector space by an inner product $\langle \cdot, \cdot \rangle$: $\|x\| = \sqrt{\langle x, x \rangle}$.
- The one-norm introduced on $\mathbb{R}^N$ by Eq. (1.5) of Sect. 1.7 (p. 11).

Norm-Induced Metric. Any norm $\| \cdot \|$ on a vector space X induces a metric d thereon via the equation: $d(x, y) := \|x - y\|$, $(x, y \in X)$. The metric d so defined is *translation-invariant:*

$$d(x + h, y + h) = d(x, y) \qquad (x, y, h \in X).$$

If X is complete in this metric it's called a *Banach space.*

Convex Sets in Normed Linear Spaces

Proposition C.5. *In any normed linear space the convex hull of a finite set of vectors is compact.*

Proof. Suppose $E := \{x_1, x_2, \ldots, x_N\}$ is our finite set of vectors. By Proposition C.4, $\operatorname{conv} E$ is the set of convex combinations of these vectors, i.e., the set of vectors $x_\lambda = \sum_1^N \lambda_j x_j$ where $\lambda = (\lambda_j)_1^N$ is a vector in the standard simplex Π_N of $\mathbb{R}^N$. Suppose $(y_n)_1^\infty$ is a sequence of vectors in $\operatorname{conv} E$. Then $y_n = x_{\eta_n}$ for some sequence $(\eta_n)_1^\infty$ of vectors in Π_N. Since Π_N is a compact subset of $\mathbb{R}^N$ we can extract a subsequence of (η_n) that converges to an element $\eta \in \Pi_N$. The corresponding subsequence of (y_n) therefore converges to $x_\eta \in \operatorname{conv} E$, thus establishing the compactness of $\operatorname{conv} E$. $\quad\square$

Recall that a subset of a metric (or topological) space is called *relatively compact* if its closure is compact. With this terminology we have the following generalization of the previous result:

Proposition C.6. *The convex hull of a relatively compact subset of a Banach space is relatively compact.*

Proof. In our Banach space X let B_r denote the open ball of radius r centered at the origin. Given subsets A and B of that space, we'll denote by $A + B$ the collection of sums $a + b$ where a ranges through A and b through B. Thanks to the completeness of X, a subset is relatively compact if and only if it is totally bounded (Proposition B.3 and Exercise B.1 of Appendix B). Thus if $A \subset X$ is totally bounded, we wish to show that $\operatorname{conv}(A)$ is totally bounded. To this end, fix $\varepsilon > 0$. Then A has an $\varepsilon/2$-net F, i.e., $A \subset F + B_{\varepsilon/2}$. Then $\operatorname{conv}(F)$, being the convex hull of a finite set, is compact, hence totally bounded, and so possesses an $\varepsilon/2$-net G, i.e., $\operatorname{conv}(F) \subset G + B_{\varepsilon/2}$. Thus

$$A \subset F + B_{\varepsilon/2} \subset \operatorname{conv}(F) + B_{\varepsilon/2}$$

and since the latter set, being the algebraic sum of two convex sets, is convex, we have $\operatorname{conv}(A) \subset \operatorname{conv}(F) + B_{\varepsilon/2}$. Putting it all together:

$$\operatorname{conv}(A) \subset \operatorname{conv}(F) + B_{\varepsilon/2} \subset G + B_{\varepsilon/2} + B_{\varepsilon/2} = G + B_\varepsilon.$$

Thus G is an ε-net for $\operatorname{conv}(A)$, so $\operatorname{conv}(A)$ is totally bounded. $\quad\square$

Remark C.7. Proposition C.6 is *not true* in the generality of normed linear spaces. For an example, in the sequence space ℓ^2 let X be the dense subspace that consists of real sequences with only finitely many non-zero terms. This is an incomplete normed linear space. Let $(e_n)_1^\infty$ be the standard orthonormal basis of ℓ^2, and for each positive integer n set $x_n = n^{-1} e_n$. Let $E = \{x_n\}_1^\infty \cup \{0\}$, a compact subset of X. However $\operatorname{conv} E$ is *not* compact; it contains each partial sum of the series $\sum_1^\infty 2^{-n} x_n$, which converges in ℓ^2 to a sum that does not belong to X. $\quad\square$

The proof of Proposition C.6 really shows that in a normed linear space, convex hulls inherit *total boundedness*. The example just presented shows that one needs to assume completeness in order to assert that convex hulls inherit *compactness*.

Operators on Normed Spaces. Let X and Y denote normed linear spaces. We'll denote the norm in either space by $\|\cdot\|$, letting context will determine the space to which the notation applies.

Proposition C.8. *Suppose X and Y are normed linear spaces and $T: X \to Y$ is a linear map. Then T is continuous on X if and only if it is bounded on some (equivalently: on every) ball therein.*

Proof. Let $B(x_0, r)$ denote the open ball in X of radius $r > 0$, with center x_0. Let $B = B(0, 1)$, the open unit ball of X. Suppose T is bounded on some ball in X, say with center x_0. Then T is bounded on some open ball $B(x_0, r)$, i.e., there exists $R > 0$ such that $T(B(x_0, r)) \subset B(0, R)$. Thus by the linearity of T:

$$RB = B(0, R) \supset T(B(x_0, r)) = T(rB + x_0) = rT(B) + Tx_0$$

hence $T(B) \subset (R/r)B - (1/r)Tx_0$ which implies, upon letting $M = \frac{R}{r} + \frac{1}{r}\|Tx_0\|$ and using the triangle inequality that $T(B) \subset MB$, i.e., that $\|Tx\| \le M$ whenever $x \in B$. If $x \in X \setminus \{0\}$ then $\xi = x/(\|x\|) \in B$, so $\|T\xi\| \le M\|\xi\|$, which translates—thanks to the linearity of T—into the inequality $\|Tx\| \le M\|x\|$, now valid for every $x \in X$. The continuity of T on X follows from this and the fact that if $x, y \in X$ then

$$\|Tx - Ty\| = \|T(x - y)\| \le M\|x - y\|.$$

Conversely, if T is continuous on X then the inverse image of the open unit ball in Y is an open subset of X that contains the origin, and so contains, for some $r > 0$, the open ball of radius r in X centered at the origin. Thus the values of T on this ball are bounded in norm by 1. By an argument similar to the one of the last paragraph, the linearity of T insures its boundedness on any ball in X.   □

Terminology. Continuous linear maps between normed spaces are often called *operators* or, thanks to the above Proposition, *bounded operators*. Here's an application of the previous results to convex sets.

Another proof of Proposition C.5. Suppose $F = \{x_1, \dots x_N\}$ is a finite set of points in the normed linear space X. Define the linear transformation $T: \mathbb{R}^N \to X$ by

$$Tx = \sum_{j=1}^{N} \xi_j x_j \text{ where } x := (\xi_1, \xi_2, \dots, \xi_N) \in \mathbb{R}^N.$$

By the Cauchy–Schwarz inequality:

$$\|Tx\| \le \sum_j |\xi_j| \|x_j\| \le M \left(\sum_j \xi_j^2 \right)^{1/2} = M\|x\|,$$

where $M = (\sum_j \|x_j\|^2)^{1/2}$. This shows that T is bounded on the (euclidean) unit ball of $\mathbb{R}^N$, and so by Proposition C.8 is continuous on $\mathbb{R}^N$.

Let C denote the convex hull of the original set F. Then $C = T(\Pi_N)$, where Π_N is the standard N-simplex. Since Π_N is compact in $\mathbb{R}^N$ and T is continuous there, C is compact. $\square$

Exercise C.1. For a normed linear space X let $\mathscr{L}(X)$ denote collection of continuous linear transformations $X \to X$.

(a)   Show that $\mathscr{L}(X)$ is an algebra under the usual algebraic operations on linear transformations.

(b)   Define the *norm* of $T \in \mathscr{L}(X)$ to be: $\|T\| := \sup\{\|Tx\| : x \in X, \|x\| \le 1\}$ ($\|T\| < \infty$ by Proposition C.8). Show that $\|\cdot\|$ is a norm on $\mathscr{L}(X)$ that makes it into a Banach space (note that it's not required here that X itself be complete).

(c)   Show that $\mathscr{L}(X)$, in the norm defined above, is a *Banach algebra*, i.e., that operator multiplication is a continuous map $\mathscr{L}(X) \times \mathscr{L}(X) \to \mathscr{L}(X)$.

C.3 Finite Dimensional Normed Linear Spaces

Here we'll work in normed linear spaces over the complex field $\mathbb{C}$. However everything we do will apply equally well to normed spaces over the reals.

Proposition C.9. *If X is a normed linear space and $\dim(X) = N < \infty$, then X is linearly homeomorphic to $\mathbb{C}^N$.*

Proof. Let $(e_j)_1^N$ be the standard unit vector basis in $\mathbb{C}^N$ (so e_j is the vector with 1 in the j-th coordinate and zeros elsewhere), and let $(x_j)_1^N$ be any basis for X. Define the map $T : \mathbb{C}^N \to X$ by $Tv = \sum_j \lambda_j(v) x_j$ where $\lambda_j(v)$ is the j-th coordinate of the vector $v \in \mathbb{C}^N$ (i.e., T is the linear map that takes e_j to x_j for $1 \le j \le N$). Thus T is a linear isomorphism taking $\mathbb{C}^N$ onto X. We've already seen in the course of proving Theorem C.5 (page 195) that T is continuous on $\mathbb{C}^N$ (actually, this proof was carried there out for $\mathbb{R}^N$, but it's the same for $\mathbb{C}^N$). Left to prove is the continuity of T^{-1}. If X were a Banach space this would follow immediately from the Open Mapping Theorem (see [103, Theorem 2.11, pp. 48–49], for example).

Here is a more elementary argument that does not require X to be complete. Let B denote the closed unit ball of $\mathbb{C}^N$ and ∂B its boundary—the unit sphere, a compact subset of $\mathbb{C}^N$ that does not contain the origin. Thus $T(\partial B)$ is, thanks to the continuity and injectivity of T, a compact subset of X that does not contain the origin. Consequently $T(\partial B)$ is disjoint from some open ball W in X that is centered at the origin.

Claim. $\Omega := T^{-1}(W)$ *is contained in* B°, *the open unit ball of* $\mathbb{C}^N$.

This will show that T^{-1} is bounded on W, hence continuous on X (Proposition C.8).

Proof of Claim. Note that:

(a) Ω is convex (thanks to the linearity of T^{-1}), hence arcwise connected.
(b) Ω contains the origin, and
(c) Ω does not intersect ∂B (thanks to the injectivity of T^{-1}).

That does it! If Ω were not contained entirely in B° its connectedness would force it to pass through ∂B, which it does not.   □

Corollary C.10. *Every finite dimensional normed linear space is complete.*

Proof. Suppose X is a finite dimensional normed linear space. By Proposition C.9, for some positive integer N there is a linear homeomorphism T taking X onto $\mathbb{C}^N$. To see that X inherits the completeness of $\mathbb{C}^N$ recall from the proof of Proposition C.8 that the continuity of T is equivalent to the existence of a positive constant M such that $\|Tx\| \le M\|x\|$ for every $x \in X$. Thus if (x_n) is a Cauchy sequence in X:

$$\|Tx_n - Tx_m\| = \|T(x_n - x_m)\| \le M\|x_n - x_m\|,$$

so the image sequence (Tx_n) is Cauchy in $\mathbb{C}^N$, hence convergent there. Thus (x_n) is the image of a convergent sequence under the continuous map T^{-1}, so it converges in X.   □

> *Warning:* It's crucial here that our homeomorphism of X onto $\mathbb{C}^N$ is *linear*. In general, a metric space homeomorphic to a complete one need not be complete. For an example, consider the arctangent function, which effects a homeomorphism of the (complete) real line onto the (incomplete) open interval $(\pi/2, \pi/2)$.

Thanks to Proposition C.9 and the exercise below: *Every linear transformation on a finite dimensional normed linear space is continuous.*

> *Exercise* C.2.   Show that every linear transformation on $\mathbb{C}^N$ is continuous.

Consequently, when working on a finite dimensional normed linear space one often uses the words "operator" and "linear transformation" interchangeably, and assumes as familiar the connection between operators and matrices.

Notes

Proposition C.9. The proof given here is taken from [103, Theorem 1.21, pp. 17–18], where it is proved in the setting of topological vector spaces. Consequently, every N dimensional subspace of a topological vector space is linearly homeomorphic to Euclidean space N-space.

Proposition C.5. The two proofs given for this result work as well for topological vector spaces.

Appendix D
Euclidean Isometries

This appendix concerns the group of isometric transformations of $\mathbb{R}^N$ with particular emphasis on the case $N = 3$, for which we'll show that the subgroup of rotations about the origin is isomorphic to the matrix group SO(3) of 3×3 real matrices with columns orthonormal in $\mathbb{R}^3$ and determinant 1.

D.1 Isometries and Orthogonal Matrices

By an *isometry* of a metric space we mean a mapping of the space into itself that preserves distances. In Euclidean space, translations and rotations are isometries. For $\mathbb{R}^N$ the isometries are easily characterized in terms of orthogonal matrices, whose definition and basic properties we'll now review.

Notation. For $\mathbb{R}^N$ we will denote the inner product by $\langle \cdot, \cdot \rangle$ and the standard unit vector basis by $(e_j : 1 \leq j \leq N)$; e_j is the vector with 1 in the j-th coordinate and zeros elsewhere. We'll think of $\mathbb{R}^N$ as a space of column vectors. For any matrix A we'll denote its transpose by A^t.

There is a fundamental connection between the inner product in $\mathbb{R}^N$ and the matrix transpose.

Proposition D.1. *For any $N \times N$ real matrix A,*

$$\langle Av, w \rangle = \langle v, A^t w \rangle \qquad (v, w \in \mathbb{R}^N).$$

Proof. It's enough to prove the result for vectors in the standard basis, so let $v = e_i$ and $w = e_j$. Then the left-hand side of the identity is just the (i, j)-element of the matrix A, while the right-hand side is, by the symmetry of the real inner product, $\langle A^t e_j, e_i \rangle$, the (j, i)-element of the transpose of A. Thus the right-hand side equals the left-hand side for these vectors hence, by the bilinearity of the inner product, for all vectors. □

© Springer International Publishing Switzerland 2016
J.H. Shapiro, *A Fixed-Point Farrago*, Universitext,
DOI 10.1007/978-3-319-27978-7

Definition D.2 (Orthogonal Matrices). To say a square matrix with real entries is *orthogonal* means that its transpose is its inverse.

More precisely: an $N \times N$ matrix A orthogonal if and only if $AA^t = A^tA = I$ where I is the $n \times n$ identity matrix. We'll use $O(N)$ to denote the collection of all $N \times N$ orthogonal matrices.

Exercise D.1. $O(N)$ is, for each positive integer N, a group under matrix multiplication.

Proposition D.3. *An $N \times N$ real matrix is orthogonal if and only if its columns form an orthonormal basis for $\mathbb{R}^N$.*

Proof. For an N-tuple of vectors in $\mathbb{R}^N$, orthonormality implies linear independence, and hence "basis-ness." Suppose A is an $n \times n$ real matrix. Its j-th column is Ae_j, so by Proposition D.1 the inner product of the j-th and k-th columns is

$$\langle Ae_j, Ae_k \rangle = \langle A^tAe_j, e_k \rangle = \text{the } (k,j)\text{-element of } A^tA$$

Thus the N-tuple of vectors $(f_j : 1 \le j \le N)$ is orthonormal if and only if $A^tA = I$. Linear algebra (or the argument above, with A replaced by A^t) shows that this happens if and only if A^t and A are inverse to each other, i.e., if and only if A is orthogonal.   □

Proposition D.4. *If A is an $N \times N$ orthogonal matrix, then:*

(a) $\langle Av, Aw \rangle = \langle v, w \rangle$ *for any pair v, w of vectors in $\mathbb{R}^N$.*
(b)   *The linear transformation $v \to Av$ is an isometry taking $\mathbb{R}^N$ onto itself.*

Proof. (a) Using successively Proposition D.1 and the definition of orthogonality:

$$\langle Av, Aw \rangle = \langle A^tAv, w \rangle = \langle v, w \rangle.$$

(b) Upon setting $v = w$ in part (a) we obtain

$$\|Av\|^2 = \langle Av, Av \rangle = \langle v, v \rangle = \|v\|^2,$$

so the mapping induced on $\mathbb{R}^N$ by A is an isometry of $\mathbb{R}^N$ into itself. Being an isometry this map is one-to-one, hence the matrix A is nonsingular, thus the induced map itself is surjective.   □

Now for the converse direction: "isometry implies linearity."

Lemma D.5. *If $T \colon \mathbb{R}^N \to \mathbb{R}^N$ is an isometry with $T(0) = 0$, then*

$$\langle T(u), T(v) \rangle = \langle u, v \rangle$$

for every pair u, v of vectors in $\mathbb{R}^N$.

Proof. This follows immediately from the relationship between norms of differences and inner products. For $u, v \in \mathbb{R}^N$:

$$\|u - v\|^2 = \langle u - v, u - v \rangle = \|u\|^2 - 2\langle u, v \rangle + \|v\|^2. \tag{D.1}$$

Upon replacing u and v in the above calculation with $T(u)$ and $T(v)$, respectively (being careful not to inadvertently assume linearity for T):

$$\begin{aligned}
\|T(u) - T(v)\|^2 &= \|T(u)\|^2 - 2\langle T(u), T(v) \rangle + \|T(v)\|^2 \\
&= \|u\|^2 - 2\langle T(u), T(v) \rangle + \|v\|^2
\end{aligned} \tag{D.2}$$

where the second equality arises from the fact that the distance from the vector 0 to v is the same as that from $0 = T(0)$ to Tv. Similarly the distance from u to v is the same as that from Tu to Tv, so the left-hand sides of Eqs. (D.1) and (D.2) are equal, hence so are the right-hand sides, and this yields the desired identity. $\qquad\square$

Proposition D.6. *If T is an isometry taking $\mathbb{R}^N$ into $\mathbb{R}^N$ with $T(0) = 0$, then there exists $A \in O(N)$ for which $T(v) = Av$ for every $v \in \mathbb{R}^N$.*

Proof. Let $(e_1, e_2, \ldots e_N)$ denote the standard orthonormal basis for $\mathbb{R}^N$. Let $f_j = Te_j$ for $1 \leq j \leq N$. Since T preserves inner products (Lemma D.5 above) there results another orthonormal basis $(f_1, f_2, \ldots f_N)$ for $\mathbb{R}^N$. Let A be the matrix that has as its j-th column the coefficients of f_j with respect to the original basis (e_j). Then $A \in O(N)$ by Proposition D.3, and $T(e_j) = Ae_j$ for each index j. Thus for every $v \in \mathbb{R}^N$:

$$T(v) = \sum_{j=1}^{N} \langle T(v), f_j \rangle f_j = \sum_{j=1}^{N} \langle T(v), T(e_j) \rangle T(e_j) = \sum_{j=1}^{N} \langle v, e_j \rangle Ae_j = Av,$$

as desired. $\qquad\square$

Theorem D.7. *A mapping $T \colon \mathbb{R}^N \to \mathbb{R}^N$ is an isometry if and only if there exists $A \in O(N)$ such that*

$$T(v) = Av + T(0) \tag{D.3}$$

for each $v \in \mathbb{R}^N$. The matrix A is uniquely determined by T.

Proof. Proposition D.4 provides one direction. For the other one note that if T is an isometry $\mathbb{R}^N \to \mathbb{R}^N$ then $T - T(0)$ is also an isometry $\mathbb{R}^N \to \mathbb{R}^N$ that additionally fixes the origin. Thus by Proposition D.6 there is an orthogonal matrix A such that $T(v) = Av + T(0)$ for each $v \in \mathbb{R}^N$. The matrix A is unique since its columns are the images of the standard basis vectors for $\mathbb{R}^N$ under the action of $T - T(0)$. $\qquad\square$

Let $\mathbb{B}^N$ denote the closed unit ball of $\mathbb{R}^N$.

Corollary D.8. *$T \colon \mathbb{B}^N \to \mathbb{B}^N$ is an isometry if and only if there exists $A \in O(N)$ such that $Tv = Av$ for every $v \in \mathbb{B}^N$.*

Proof. The proof of Theorem D.7 actually showed that:

An isometry $T \colon \mathbb{B}^N \to \mathbb{R}^N$ must have the form (D.3) for each $v \in \mathbb{B}^N$.

Suppose, in addition, that T maps $\mathbb{B}^N$ into itself. It seems obvious that the translation vector $T(0)$ must then equal 0; for a picture-free argument let's suppose this is not the case. Let $x_0 = T(0)$ and $u = A^{-1}(x_0/\|x_0\|)$. Since $A^{-1} = A^t$ is also an orthogonal matrix, u is a unit vector (Proposition D.4), so belongs to $\mathbb{B}^N$. However

$$Tu = \frac{x_0}{\|x_0\|} + x_0 = (1 + \|x_0\|)\frac{x_0}{\|x_0\|},$$

so $\|Tu\| > 1$, contradicting our assumption that $T(\mathbb{B}^N) \subset \mathbb{B}^N$. Thus $x_0 = 0$. $\quad\square$

Corollary D.9. *Isometries $\mathbb{R}^N \to \mathbb{R}^N$ and $\mathbb{B}^N \to \mathbb{B}^N$ must be surjective.*

Corollary D.10. *For $N \geq 2$ the isometry groups of $\mathbb{R}^N$ and $\mathbb{B}^N$ are not commutative.*

Proof. In view of Theorem D.7 and Corollary D.8 it's enough to note that:

If $N \geq 2$ then the matrix group $O(N)$ is not commutative.

Indeed, here are two matrices in $O(2)$ that do not commute:

$$\begin{pmatrix} \frac{1}{\sqrt{2}} & -\frac{1}{\sqrt{2}} \\ \frac{1}{\sqrt{2}} & \frac{1}{\sqrt{2}} \end{pmatrix} \quad \text{and} \quad \begin{pmatrix} 1 & 0 \\ 0 & -1 \end{pmatrix}$$

the first of which induces rotation through an angle of 45 degrees, while the second induces reflection about the horizontal axis. To get an example in $O(N)$ for $N > 2$ just put each of the above matrices in the upper left-hand corner of an $N \times N$ matrix, and fill in the remaining entries with ones on the main diagonal and zeros off it. $\quad\square$

D.2 Rotations of $\mathbb{R}^2$ and $\mathbb{R}^3$

Every orthogonal matrix has determinant ± 1; those with determinant 1 are called *special-orthogonal*. The special-orthogonal matrices of a given size form a subgroup of all the invertible matrices of that size. Here we'll be concerned with $SO(2)$ and $SO(3)$, the special-orthogonal matrices respectively of sizes 2×2 and 3×3.

Proposition D.11. *Each matrix in $SO(2)$ induces on $\mathbb{R}^2$ a rotation about the origin. If a matrix in $O(2)$ has determinant -1, then it induces on $\mathbb{R}^2$ a reflection in a line through the origin.*

Proof. Each $A \in O(2)$ takes the pair of unit vectors (e_1, e_2) (respectively along the horizontal and vertical axes) to an orthogonal pair (u, v) of unit vectors, where u is the rotate of e_1 through some angle θ, and v is either the rotate of e_2 through that angle—in which case the determinant of A is 1 and A is the mapping of "rotation by θ"—or v is the negative of that vector. In this latter case $\det A = -1$, and A effects the mapping of reflection in the line through the origin parallel to u. $\qquad\square$

Proposition D.12. *If $A \in \mathrm{SO}(3)$ then the map $x \to Ax$ is a rotation of $\mathbb{R}^3$, with center at the origin.*

We're saying that for each $A \in \mathrm{SO}(3)$ the associated linear transformation fixes a line through the origin, and acts as a rotation about this line (the so-called axis of rotation).

Proof. Suppose $A \in \mathrm{SO}(3)$. To find the axis of rotation we need to show that $Av = v$ for some unit vector $v \in \mathbb{R}^3$, i.e., that 1 is an eigenvalue of A, or equivalently that $\det(A - I) = 0$. For this, note that since $AA^t = I$ we have

$$(A - I)A^t = AA^t - A^t = I - A^t = -(A - I)^t$$

hence, since $\det A = \det A^t = 1$:

$$\det(A - I) = \det(A - I)\det(A^t) = \det[(A - I)A^t]$$

$$= \det[-(A - I)^t] = (-1)^3 \det(A - I)^t$$

$$= -\det(A - I)$$

so $\det(A - I) = 0$, as desired.

Let $v_1 \in \mathbb{R}^3$ be the unit vector promised by the last paragraph: $Av_1 = v_1$. Let (v_2, v_3) be an orthonormal basis for the subspace E of $\mathbb{R}^3$ orthogonal to v_1. Then (v_1, v_2, v_3) is an orthonormal basis for $\mathbb{R}^3$, relative to which the matrix of the transformation $x \to Ax$ has block diagonal form

$$\begin{bmatrix} 1 & 0 \\ 0 & B \end{bmatrix} \tag{D.4}$$

where B is a 2×2 orthogonal matrix. Thus A and B have the same determinant, so $\det B = 1$, i.e., $B \in \mathrm{SO}(2)$, so by the previous proposition B induces on E either the identity map or a rotation about the origin. $\qquad\square$

Corollary D.13. *A map $T \colon \mathbb{R}^3 \to \mathbb{R}^3$ is a rotation about the origin if and only if there exists a matrix $A \in \mathrm{SO}(3)$ such that $T(v) = Av$ for every $v \in \mathbb{R}^3$.*

Proof. We already know (Proposition D.12) that maps of $\mathbb{R}^3$ represented by matrices in $\mathrm{SO}(3)$ are rotations about the origin. For the converse, suppose T is a rotation of $\mathbb{R}^3$ about the origin, i.e., an isometry of $\mathbb{R}^3$ that fixes a line through the origin about which it acts as a two dimensional rotation. By Corollary D.8 we know that T is represented as left multiplication by an orthogonal matrix A. Thus A must have the block-diagonal form (D.4) with $B \in \mathrm{SO}(2)$, i.e., $A \in \mathrm{SO}(3)$. $\qquad\square$

It's easy to see that the rotation group of the ball does *not* share the commutativity of that of the disc; take, for an example, a pair of 45° rotations about different orthogonal axes. Thus, while the matrix group SO(2) is commutative, SO(3) is not.

Exercise D.2. SO(N) is not commutative for every $N \geq 3$.

The matrix of a rotation in space. Rotations about the origin in three-space are linear transformations, and linear transformations have matrix representations. Let $R_u(\rho)$ denote the matrix (with respect to the standard basis of $\mathbb{R}^3$) of the transformation of rotation about the origin through angle ρ with axis the unit vector $u \in \mathbb{R}^3$ (the "right-hand rule" determining the positive direction of ρ). Although somewhat complicated, this matrix factors readily as a product of simpler matrices. We start with the three "elementary" rotation matrices; the ones that represent rotations about the coordinate axes:

1. Rotation through angle ρ about the z-axis

$$R_z := \begin{bmatrix} \cos\rho & -\sin\rho & 0 \\ \sin\rho & \cos\rho & 0 \\ 0 & 0 & 1 \end{bmatrix}$$

2. Rotation through angle ρ about the x-axis

$$R_x := \begin{bmatrix} 1 & 0 & 0 \\ 0 & \cos\rho & -\sin\rho \\ 0 & \sin\rho & \cos\rho \end{bmatrix}$$

3. Rotation through angle ρ about the y-axis

$$R_y := \begin{bmatrix} \cos\rho & 0 & \sin\rho \\ 0 & 1 & 0 \\ -\sin\rho & 0 & \cos\rho \end{bmatrix}$$

Now fix the unit vector $u \in \mathbb{R}^3$ and the angle $\rho \in [-\pi, \pi)$, and let L_u denote the oriented line through the origin in the direction of u. We're going to understand the transformation $R_u(\rho)$ of rotation about L_u through angle ρ by factoring it into a product of several elementary ones. For this let (φ, θ) be the spherical coordinates of u, i.e.,

$$u = [\sin\varphi\cos\theta, \sin\varphi\sin\theta, \cos\varphi]^t$$

where $\varphi \in [0, \pi]$ is the angle between u and the z-axis, and $\theta \in [-\pi, \pi]$ is the angle between the x-axis and the projection of the u into the x, y-plane.

Let $T = R_y(-\varphi)R_z(-\theta)$, so that T rotates u through angle $-\theta$ about the z-axis, depositing it into the x,z-plane, then in that plane (i.e., about the y-axis) rotates the resulting vector through angle ρ so that it ends up at the "north pole" $e_3 := [0,0,1]^t$. Thus $T^{-1}R_z(\rho)T$ fixes u and, since T belongs to SO(3), and therefore preserves both distances and angles, it rotates points of $\mathbb{R}^3$ about L_u through angle ρ, i.e., it's none other than $R_u(\rho)$. Explicitly:

$$R_u(\rho) = R_z(\theta)R_y(\varphi)R_z(\rho)R_y(-\varphi)R_z(-\theta). \tag{D.5}$$

To find the matrix of $R_u(\rho)$ (with respect to the standard unit vector basis of $\mathbb{R}^3$) one "need only" multiply the elementary matrices for the five transformations on the right-hand side of (D.5). This is best done with your favorite computer-algebra program; the result is nevertheless quite a mess. To bring it into some kind of reasonable form it helps to invert the spherical coordinate representation of u, noting that $\cos\varphi = z$, $\sin\varphi = \sqrt{x^2+y^2} = \sqrt{1-z^2}$ (non-negative square root because $0 \le \varphi \le \pi$), $\cos\theta = x/\sqrt{1-z^2}$, and $\sin\theta = y/\sqrt{1-z^2}$.

Again with the help of your computer-algebra program, most likely aided by some paper and pencil algebraic simplifications, there will result the following matrix representation of $R_u(\rho)$:

$$\begin{bmatrix} x^2 + (1-x^2)\cos\rho & xy(1-\cos\rho)-z\sin\rho & xz(1-\cos\rho)+y\sin\rho \\ xy(1-\cos\rho)+z\sin\rho & y^2+(1-y^2)\cos\rho & yz(1-\cos\rho)-x\sin\rho \\ xz(1-\cos\rho)-y\sin\rho & yz(1-\cos\rho)+x\sin\rho & z^2+(1-z^2)\cos\rho \end{bmatrix}.$$

Notes

Rotations in $\mathbb{R}^3$ and beyond. Proposition D.12 (or more accurately, the statement that each rotation in $\mathbb{R}^3$ about the origin has a fixed axis), was first proved by Euler in 1775–1776; it's called "Euler's Rotation Theorem." For a lively article that gives much more detail about this result, see [90]. The results above on SO(2) and SO(3) generalize to higher dimensions, but now reflections can be present. For O(N) the full story is this (see, e.g., [14, Theorem 10.12, p. 152]):

For $A \in$ O(N) there exists an orthonormal basis for $\mathbb{R}^N$ relative to which the transformation $x \to Ax$ has block diagonal matrix $(I_p, -I_q, B_1, \ldots, B_r)$ where the I's are identity matrices of orders p and q respectively, the B's are 2×2 orthogonal matrices, and $p+q+2r = n$.

Appendix E
A Little Group Theory, a Little Set Theory

We'll write groups multiplicatively, denoting the identity element by "e". For subsets A and B of a group G we'll write AB for the collection of all products ab with $a \in A$ and $b \in B$, using the abbreviation aB for the product $\{a\}B$. If H is a subgroup of G (i.e., a group in the operation inherited from G, whose identity is the identity element of G), and $g \in G$, then gH is the *left coset of G modulo H,* and Hg is the corresponding *right coset.*

E.1 Normal Subgroups

Definition E.1 (Normal Subgroup). Suppose G is a group and H a subgroup (notation: $H < G$). To say H is a *normal* subgroup of G (notation: $H \lhd G$) means that $gH = Hg$ for any $g \in G$, i.e., there is no distinction between left and right cosets. In this case we'll use G/H to denote the collection of all cosets of G modulo H.

Proposition E.2. *Suppose H is a subgroup of G. Then $H \lhd G$ if and only if G/H forms a group under the multiplication*

$$(g_1 H)(g_2 H) = g_1 g_2 H \qquad (g_1, g_2 \in G).$$

Proof. If $H \lhd G$ and $g_1, g_2 \in G$, then

$$(g_1 H)(g_2 H) = g_1 (Hg_2)H = g_1 g_2 HH = g_1 g_2 H$$

and from this we have for each $g \in G$:

$$(gH)(g^{-1}H) = gg^{-1}H = eH = H.$$

Thus G/H is a group under the inherited multiplication, with H being the identity and $(gH)^{-1} = g^{-1}H$.

© Springer International Publishing Switzerland 2016
J.H. Shapiro, *A Fixed-Point Farrago*, Universitext,
DOI 10.1007/978-3-319-27978-7

Conversely, if G/H is a group under the multiplication $(g_1H)(g_2H) = g_1g_2H$ (where it's being assumed that the multiplication is well defined), then H is the identity, and for any $g \in G$:

$$gHg^{-1} \subset gHg^{-1}H = gg^{-1}H = eH = H$$

so $gH \subset Hg$. For the opposite inclusion just replace g by its inverse in this one and take inverses of both sides of the resulting inclusion.   □

E.2 Solvable Groups

Let G be a group. For each pair a,b of elements of G let $[a,b]$ denote the commutator $a^{-1}b^{-1}ab$ (so named because if $c = [a,b]$ then $ab = bac$). To think about solvability for G let's consider *chains of subgroups*

$$\{e\} = G_0 < G_1 < \ldots < G_n = G. \tag{E.1}$$

Here's a restatement of Definition 12.2(c) of "solvable group."

Definition E.3. To say that G is a *solvable group* means that there is a chain of subgroups (E.1) with each subgroup G_k containing all the commutators of G_{k+1} ($0 \leq k < n$).

Good things happen whenever a subgroup contains all the commutators of its parent group.

Proposition E.4. *Suppose G is a group and H a subgroup of G. Then the following are equivalent:*

(a) H contains all the commutators of G.
(b) H is a normal subgroup of G and the quotient group G/H is abelian.

Proof. (a) → (b): Suppose H contains all the commutators of G, i.e., for every pair a,b of elements of G there is an element $h \in H$ (namely $h = [a,b] = a^{-1}b^{-1}ab$) such that $ab = bah$. In particular, for any $a \in G$ and $h_1 \in H$ there exists $h \in H$ such that

$$a^{-1}(h_1a) = a^{-1}(ah_1h) = h_1h \in H.$$

Thus for each $a \in G$ we have $a^{-1}Ha \subset H$, so $Ha \subset aH$, hence—as we've seen above (last part of proof of Proposition E.2)—this implies $Ha = aH$, i.e., H is a normal subgroup of G. As for the commutativity of the quotient group G/H, note that if $a,b \in G$ then, as noted above, $ab = bah$ for $h = [a,b] \in H$ hence $abH \subset baH$ and so $abH = baH$.

(b) → (a): Consider the statement

(*) $H < G$ and $abH = baH$ for each pair a,b of elements of G.

Then for all $a, b \in G$ and $h_1 \in H$ there exists $h_2 \in H$ such that $abh_1 = bah_2$, i.e., $[a, b] = h_2 h_1^{-1} \in H$. Thus statement (*) implies that each commutator of G belongs to H, which by the first part of our proof is enough to guarantee normality for H, and hence—again by (*)—commutativity for G/H.   □

E.3 The Axiom of Choice and Zorn's Lemma

If we are given a family of sets, the Axiom of Choice allows us to choose one element from each member of the family. More precisely:

The Axiom of Choice. *Suppose X is a set and $\mathscr{E}$ is a family of nonempty subsets of X. Then there is a "choice function" $f: \mathscr{E} \to X$ such that $f(E) \in E$ for each $E \in \mathscr{E}$.*

Definition. A *partial order* on a set X is a binary relation "$\leq$" such that for all $x, y, z \in X$:

$x \leq x$ (reflexivity),
$x \leq y$ and $y \leq x \implies x = y$ (antisymmetry), and
$x \leq y$ and $y \leq z \implies x \leq z$ (transitivity).

Suppose "$\leq$" is a partial order on X and $S \subset X$. An element $b \in X$ for which each $s \in S$ is $\leq b$ is called an *upper bound* for S. An element $m \in X$ is called *maximal* if no other element of X "exceeds" m, i.e., if $x \in X$ and $m \leq x$ then $x = m$. $S \subset X$ is said to be *totally ordered* if for every pair of elements $s, t \in S$ either $s \leq t$ or $t \leq s$.

Zorn's Lemma. *If X is a partially ordered set in which every totally ordered subset has an upper bound, then X has a maximal element.*

Zorn's Lemma and the Axiom of Choice are *equivalent* in the sense that each one can be derived from the other. Halmos [46, Sects. 15 and 16] gives an delightful introduction to these two principles of mathematics, with a proof of their equivalence.

Notes

Non-normal pathology. The following exercise, taken from taken from Milne's freely downloadable introduction to group theory [77], shows that we can't expect to extend the definition used to multiply cosets modulo normal subgroups to cosets modulo arbitrary subgroups.

> *Exercise E.1.* Let G denote the collection of all 2×2 invertible matrices with rational entries, and let H denote the set of 2×2 matrices of the form $\left[\begin{smallmatrix} 1 & n \\ 0 & 1 \end{smallmatrix}\right]$ where n runs through the integers. Then G is a group under matrix multiplication and H is a subgroup (isomorphic to the group of integers under addition).

(a)   Show that H is not a normal subgroup of G. In fact, if $g = \begin{bmatrix} 5 & 0 \\ 0 & 1 \end{bmatrix}$ then $g \in G$ and gH
is a proper subset of Hg. Moreover, $g^{-1}H$ is not even contained in Hg^{-1}.

(b)   Show that for g as in part (a), $g^{-1}HgH \neq H$.

The usual definition of "solvable." Proposition E.4 shows that our definition of "solvable group" can be restated

G is a solvable group if and only if there exists a chain (E.1) *of normal subgroups such that each group* G_k/G_{k-1} *is commutative.*

This is the usual definition of "solvable" for groups.

The Axiom of Choice. Consequences such as the existence of non-measurable subsets of the real line, and more spectacularly the Banach-Tarski Paradox, initially gave the Axiom of Choice something of a bad reputation. To quote Halmos [46, Sect. 15, p. 60]:

> It used to be considered important to examine, for each consequence of the axiom of choice, the extent to which the axiom is needed in the proof of the consequence. An alternative proof without the axiom of choice spelled victory; a converse proof, showing that the consequence is equivalent to the axiom . . . meant honorable defeat. Anything in between was considered exasperating.

The Axiom of Choice is now much better understood. In Chap. 13 of Wagon's book [121] there is a detailed discussion of the role it plays in set theory. On page 214 (Fig. 13.1) of that chapter there is a useful diagram showing the logical connections between the Axiom of Choice and various well-known theorems of mathematics that follow from it (e.g., the Tychonoff Product Theorem, the Hahn-Banach Theorem). Jech's monograph [55] provides an in-depth treatment of such matters.

References

1. Agnew, R.P., Morse, A.P.: Extensions of linear functionals, with applications to limits, integrals, measures, and densities. Ann. of Math (2), **39**, 20–30 (1938)
2. Apostol, T.M.: Mathematical Analysis, A Modern Approach to Advanced Calculus. Addison-Wesley, Reading (1957)
3. Argyros, S.A., Haydon, R.G.: A hereditarily indecomposable $\mathscr{L}_\infty$-space that solves the scalar-plus-compact problem. Acta Math. **206**, 1–54 (2011)
4. Aronszajn, N., Smith, K.T.: Invariant subspaces of completely continuous operators. Ann. of Math (2), **60**, 345–350 (1954)
5. Banach, S.: Sur les opérations dans les ensembles abstraits et leur application aux équations intégrales. Fund. Math. **3**, 133–181 (1922)
6. Banach, S.: Sur le probleme de la mesure. Fund. Math. **4**, 7–33 (1923)
7. Banach, S.: Un théorème sur les transformations biunivoques. Fund. Math. **6**, 236–239 (1924)
8. Banach, S.: On Haar's measure. In: Theory of the Integral, by Stanisław Saks, Monografie Matematyczne, Warsaw, vol. 7, pp. 314–319 (1937) Reprinted by Dover Pub. 1964 & 2005
9. Banach, S.: Théorie des Opérations Linéaires, 2nd edn. Chelsea (1978). First ed. published in 1932, Monografje Matematyczne, vol. I, Warsaw
10. Banach, S., Tarski, A.: Sur la décomposition des ensembles de points en parties respectivement congruentes. Fund. Math. **6**, 244–277 (1924)
11. Bernstein, A.R., Robinson, A.: Solution of an invariant subspace problem of K. T. Smith and P. R. Halmos. Pacific J. Math. **16**, 421–431 (1966)
12. Bilton, N., Sandhaus, E.: # of links on the homepage of 98 top websites. Available at http://www.nickbilton.com/98/
13. Birkhoff, G.: A note on topological groups. Compos. Math. **3**, 427–430 (1936)
14. Blyth, T., Robertson, E.F.: Further Linear Algebra. Springer, New York (2006). Second printing
15. Boyce, W.M.: Commuting functions with no common fixed point. Trans. Amer. Math. Soc. **137**, 77–92 (1969)
16. Bressan, A.: Noncooperative Differential Games. A Tutorial. Lecture Notes available online at http://www.math.psu.edu/bressan/PSPDF/game-lnew.pdf (2010)
17. Brin, S., Page, L.: The anatomy of a large-scale hypertextual web search engine. In: Seventh International World-Wide Web Conference (WWW 1998), Brisbane, 14–18 April 1998. Available online at http://infolab.stanford. edu/pub/papers/google.pdf (1998)
18. Brouwer, L.E.J.: Über Abbildung von Mannigfaltigkeiten. Math. Ann. **71**, 97–115 (1912)
19. Brouwer, L.E.J.: An intuitionist correction of the fixed-point theorem on the sphere. Proc. R. Soc. Lond. Ser. A Math. Phys. Eng. Sci. **213**, 1–2 (1952)

© Springer International Publishing Switzerland 2016
J.H. Shapiro, *A Fixed-Point Farrago*, Universitext,
DOI 10.1007/978-3-319-27978-7

20. Bryan, K., Leise, T.: The $25,000,000,000 eigenvector: the linear algebra behind Google. SIAM Rev. **48**, 569–581 (2006)
21. Casti, J.L.: Five Golden Rules: Great Theories of 20th Century Mathematics, and Why They Matter. Wiley, New York (1965)
22. Cauty, R.: Solution du problème de point fixe de Schauder. Fund. Math. **170**, 231–246 (2001)
23. Cellina, A.: Approximation of set valued functions and fixed point theorems. Ann. Mat. Pura Appl. **82**(4), 17–24 (1969)
24. Chalendar, I., Partington, J.R.: Modern Approaches to the Invariant-Subspace Problem. Cambridge University Press, Cambridge (2011)
25. Chernoff, P.R.: A simple proof of Tychonoff's theorem via nets. Amer. Math. Monthly **99**, 932–934 (1992)
26. Cohen, D.I.A.: On the Sperner Lemma. J. Comben. Theory **2**, 585–587 (1967)
27. Day, M.M.: Means for the bounded functions and ergodicity of the bounded representations of semi-groups. Trans. Amer. Math. Soc. **69**, 276–291 (1950)
28. Day, M.M.: Amenable semigroups. Illinois J. Math. **1**, 509–544 (1957)
29. Diestel, J., Spalsbury, A.: The Joys of Haar Measure. Graduate Studies in Mathematics, vol. 150. American Mathematical Society, Providence (2014)
30. Ding, J., Zhou, A.: Non-Negative Matrices, Positive Operators, and Applications. World Scientific, Singapore (2009)
31. Dobrowolski, T.: Revisiting Cauty's proof of the Schauder conjecture. Abstr. Appl. Anal. **7**, 407–433 (2003)
32. Drinfel'd, V.G.: Finitely additive measures on S^2 and S^3, invariant with respect to rotations. Funct. Anal. Appl. **18**, 245–246 (2014)
33. Driver, R.D.: Torricelli's law-an ideal example of an elementary ODE. Amer. Math. Monthly **105**, 453–455 (1998)
34. Dugundji, J.: An extension of Tietze's theorem. Pacific. J. Math. **1**, 353–367 (1951)
35. Dugundji, J., Granas, A.: A proof of the Ryll-Nardzewski fixed point theorem. J. Math. Anal. Appl. **97**, 301–305 (1983)
36. Dym, H., McKean, H.P.: Fourier Series and Integrals. Academic, New York (1972)
37. Edwards, R.E.: Functional Analysis: Theory and Applications. Holt, Rinehart, Winston, New York (1965)
38. Enflo, P.: On the invariant subspace problem for Banach spaces. Acta Math. **158**, 213–313 (1987)
39. Folland, G.B.: A Course in Abstract Harmonic Analysis. CRC, Boca Raton (1995)
40. Franklin, J.: Methods of Mathematical Economics. Springer, Heidelberg (1980)
41. Galilei, G.: Dialogues Concerning Two New Sciences. Dover, New York (1914). Translation by Crew and de Salvio of: Discorsi e Dimonstrazione Matematiche Intorno à Due Nuove Scienze. Elzevir, Leiden (1638). This translation originally published by Macmillan, New York (1914)
42. Haar, A.: Der Masbegriff in der Theorie der Kontinuierlichen Gruppen. Ann. of Math (2), **34**, 147–169 (1933)
43. Hadlock, C.R.: Field Theory and its Classical Problems. Carus Monograph, vol. 19. Mathematical Association of America, Washington (1978)
44. Hadwin, D.W., Nordgren, E.A., Radjavi, H., Rosenthal, P.: An operator not satisfying Lomonosov's hypotheses. J. Funct. Anal. **38**, 410–415 (1980)
45. Hahn, F.: A fixed-point theorem. Math. Syst. Theory **1**, 55–57 (1968)
46. Halmos, P.R.: Naive Set Theory. van Nostrand, Princeton (1960); Undergraduate Texts in Mathematics. Reprinted by Springer, New York (1974)
47. Hardin, G.: The tragedy of the commons. Science **162**, 1243–1248 (1968)
48. Hausdorff, F.: Grundzüge der Mengenlehre. Veig, Leipzig (1914)
49. Hawkins, T.: Weyl and the topology of continuous groups. In: James, I.M. (ed.) History of Topology. North Holland, Amsterdam (1999)
50. Hiriart-Urrutz, J.B., Lemaréchal, C.: Convex Analysis and Minimization Algorithms I. Springer, Heidelberg (1996)

51. Horn, R.A., Johnson, C.: Matrix Analysis, 2nd edn. Cambridge University Press, New York (2013)
52. Huneke, J.P.: On common fixed points of commuting continuous functions on an interval. Proc. Amer. Math. Soc. **139**, 371–381 (1969)
53. Ivanov, N.V.: A differential forms perspective on the Lax proof of the change of variable formula. Amer. Math. Monthly **112**, 799–806 (2005)
54. Izzo, A.J.: A functional analysis proof of the existence of Haar measure on locally compact abelian groups. Proc. Amer. Math. Spc. **115**, 581–583 (1992)
55. Jech, T.J.: The Axiom of Choice. North Holland, Amsterdam (1973). Republished by Dover, New York (2008)
56. Kakutani, S.: Two fixed-point theorems concerning bicompact convex sets. Proc. Imp. Acad. (Jap.) **14**, 242–245 (1938)
57. Kakutani, S.: Uber die Metrisation der topologischen Gruppen. Proc. Imp. Acad. (Jap.) **14**, 82–84 (1938)
58. Kakutani, S.: A generalization of Brouwer's fixed point theorem. Duke Math. J. **8**, 457–459 (1941)
59. Kakutani, S.: Topological properties of the unit sphere of a Hilbert space. Proc. Imp. Acad. Tokyo **19**, 269–271 (1943)
60. Kelley, J.L., Namioka, I.: Linear Topological Spaces. van Nostrand, New York (1963)
61. Klarreich, E.: The Mathematics of Strategy. Article available online from http://www.pnas.org/site/classics/classics5.xhtml
62. Klee, V.L.: Some topological properties of convex sets. Trans. Amer. Math. Soc. **78**, 30–45 (1955)
63. Knaster, B.: Un théorème sur les fonctions d'ensembles. Ann. Soc. Polon. Math. **6**, 133–134 (1928)
64. Kuhn, H.W.: Simplicial approximation of fixed points. Proc. Natl. Acad. Sci. U.S.A. **61**, 1238–1242 (1968)
65. Kuhn, H.W., Tucker, A.W.: John von Neumann's work in the theory of games and mathematical economics. Bull. Am. Math. Soc. **64**, 100–122 (1958)
66. Lalley, S.: Markov Chains: Basic Theory. Lecture Notes available from http://galton.uchicago.edu/ lalley/Courses/312/MarkovChains.pdf (2009)
67. Lax, P.D.: Change of variables in multiple integrals. Amer. Math. Monthly **106**, 497–501 (1999)
68. Lemke, C.E.: Bimatrix equilibrium points in mathematical programming. Managment Sci. **11**, 681–689 (1965)
69. Li, M.C.: An elementary proof of a generalization of Banach's Mapping Theorem. Amer. Math. Monthly **121**, 445–446 (2014)
70. Lindelöf, M.E.: Sur l'application de la méthode des approximations successives aux équations différentiales ordinaires du premier ordre. C. R. Math. Acad. Sci. Paris **114**, 454–457 (1894)
71. Lomonosov, V.I.: Invariant subspaces for the family of operators which commute with a completely continuous operator. Funct. Anal. Appl. **7**, 213–214 (1973)
72. MacCluer, C.R.: The many proofs and applications of Perron's Theorem. SIAM Rev. **42**, 487–498 (2000)
73. Magnus, W., Karrass, A., Solitar, D.: Combinatorial Group Theory: Presentations of Groups in Terms of Generators and Relations, 2nd edn. Dover, New York (1976). Originally published by Interscience Publishers, New York (1966)
74. Margulis, G.: Some remarks on invariant means. Monatsh. Math. **90**, 233–235 (1980)
75. Markov, A.A.: Quelques théoremes sur les ensembles abéliens. C.R. URSS **2**, p. 311 (1936)
76. Mason, O.: NonNegative Matrices and Related Topics. Lecture notes available at www.hamilton.ie/ollie/Downloads/Pos1.pdf
77. Milne, J.S.: Group Theory (v3.13). Available at www.jmilne.org/math/ (2013)
78. Milnor, J.: Analytic proofs of the "hairy ball theorem" and the Brouwer fixed point theorem. Amer. Math. Monthly **85**, 521–524 (1978)

79. Milnor, J.: A Nobel Prize for John Nash. Math. Intelligence **17**, 11–17 (1995)
80. Montgomery, D., Zippin, L.: Topological Transformation Groups. Interscience, New York (1955)
81. Myerson, R.B.: Nash equilibrium and the history of economic theory. J. Econ. Lit. **37**, 1067–1082 (1999)
82. Namioka, I., Asplund, E.: A geometric proof of Ryll-Nardzewski's fixed point theorem. Bull. Amer. Math. Soc. **73**, 443–445 (1967)
83. Nasar, S.: The lost years of a Nobel laureate. The N. Y. Times, Nov 13, 1994
84. Nasar, S.: A Beautiful Mind. Simon & Schuster, New York (1998)
85. Nash Jr., J.F.: Equilibrium points in N-person games. Proc. Natl. Acad. Sci. U.S.A. **36**, 48–49 (1950)
86. Nash Jr., J.F.: Non-cooperative games, pp. 1–32. Dissertation, Princeton (1950)
87. Nash Jr., J.F.: Non-cooperative games. Ann. of Math (2), **54**, 286–295 (1951)
88. von Neumann, J.L.: Zur algemeinnen theorie des masses. Fund. Math. **13**, 73–116 (1929)
89. von Neumann, J.L.: A model of general economic equilibrium. Rev. Econ. Studies **13**, 1–9 (1945)
90. Palais, B., Palais, R., Rodi, S.: A disorienting look at Euler's theorem on the axis of a rotation. Amer. Math. Monthly **116**, 892–909 (2009)
91. Park, S.: Ninety years of the Brouwer fixed point theorem. Vietnam J. Math. **27**, 187–222 (1999)
92. Pearcy, C., Shields, A.L.: A survey of the Lomonosov technique. Topics in Operator Theory, Mathematical Surveys, vol. 13, pp. 221–229. American Mathematical Society, Providence (1974)
93. Perko, L.: Differential Equations and Dynamical Systems, 3rd edn. Springer, New York (2000)
94. Radjavi, H., Rosenthal, P.: Invariant Subspaces, 2nd edn. Dover, New York (2003). First edition published by Springer, New York (1973)
95. Read, C.J.: A short proof concerning the invariant subspace problem. J. Lond. Math. Soc. (2) **34**, 335–348 (1986)
96. Read, C.J.: The invariant subspace problem for a class of Banach spaces 2: hypercyclic operators. Isreal J. Math. **63**, 1–40 (1988)
97. Riesz, F.: Sur les opérations fonctionelles linéaires. C. R. Math. Acad. Sci. Paris **149**, 974–977 (1909)
98. Riesz, F., Nagy, B.S.: Functional Analysis, 2nd edn. Frederick Ungar, New York (1955). Reprinted by Dover Publications, New York (1990)
99. Rogers, C.A.: A less strange version of Milnor's proof of Brouwer's fixed-point theorem. Amer. Math. Monthly **67**, 525–527 (2010)
100. Rudin, W.: Fourier Analysis on Groups. Interscience, vol. 12 Wiley, New York (1962)
101. Rudin, W.: Principles of Mathematical Analysis, 3rd edn. McGraw-Hill, New York (1976)
102. Rudin, W.: Real and Complex Analysis, 3rd edn. McGraw-Hill, New York (1987)
103. Rudin, W.: Functional Analysis, 2nd edn. McGraw-Hill, New York (1991)
104. Runde, V.: Lectures on Amenability. Springer Lecture Notes, vol. 1774. Springer, Berlin (2002)
105. Ryll-Nardzewski, C.: On fixed points of semigroups of endomorphisms of linear spaces. In: Proceeding of the Fifth Berkeley Sympos Math. Statistics & Probability, pp. 55–61 (1967)
106. Scarf, H.E.: Fixed-point theorems and economic analysis: mathematical theorems can be used to predict the probable effects of changes in economic policy. Am. Sci. **71**, 289–296 (1983)
107. Schauder, J.: Der Fixpunktsatz in Funktionalräumen. Studia. Math. **2**, 171–180 (1930)
108. Schilling, F., Piper, W.S.: Basic Abstract Algebra. Allyn & Bacon, Boston (1975)
109. Schur, I.: Über die charakteristischen Wurzeln einerlinearen Substitution mit einer Anwendung auf die Theorie der Integralgleichungen. Math. Ann. **66**, 488–510 (1909)
110. Shurman, J.: Multivariable Calculus. Lecture Notes for a Course at Reed College. Available online at http://people.reed.edu/ jerry/211/vcalc.pdf

111. Sperner, E.: Neur Beweis für die Invarianz der Dimensionzal und des Gebietes. Abh. Math. Sem. Univ. Hamburg **6**, 265–272 (1928)
112. Stuckless, T.: Brouwer's Fixed Point Theorem: Methods of Proof and Generalizations. Available from http://oldweb.cecm.sfu.ca/Preprints03/ preprints03.html (2003)
113. Su, F.E.: Rental harmony: Sperner's lemma in fair division. Amer. Math. Monthly **106**, 930–942 (1999)
114. Sullivan, D.: For $n > 3$ there is only one finitely additive rotationally invariant measure on the n-sphere defined on all Lebesgue measurable subsets. Bull. Amer. Math. Soc. **4**, 121–123 (1981)
115. Tao, T.: The Banach-Tarski Paradox. Available online from www.math.ucla.edu/~tao/preprints/Expository/banach-tarski.pdf
116. Tarski, A.: A lattice-theoretical fixpoint theorem and its applications. Pacific J. Math. **5**, 255–309 (1955)
117. Trefethen, L.N., Bau III, D.: Numerical Linear Algebra. SIAM, Philadelphia (1997)
118. Trench, W.F.: Introduction to Real Analysis. Freely available online at http://ramanujan.math.trinity.edu/wtrench/texts/TRENCH_REAL_ANALYSIS .PDF (2013)
119. Troitsky, V.G.: Lomonosov's theorem cannot be extended to chains of four operators. Proc. Amer. Math. Soc. **128**, 521–525 (2000)
120. Tychonoff, A.: Ein Fixpunktsatz. Math. Ann. **111**, 767–776 (1935)
121. Wagon, S.: The Banach-Tarski Paradox, paperback edn. Cambridge University Press, Cambridge (1993)
122. Wapner, L.: The Pea and the Sun: A Mathematical Paradox. A. K. Peters Ltd., Wellesley (2005)
123. Willard, S.: General Topology. Addison Wesley, Reading (1970). Reprinted by Dover Publications, Mineola (2004)
124. Williamson, J.H.: Compact linear operators in linear topological spaces. J. Lond. Math. Soc. (2) **29**, 149–156 (1954)
125. Young, N.: An Introduction to Hilbert Space. Cambridge University Press, Cambridge (1998)

Index

List of symbols

A^t The transpose of the matrix A. 199

$B(S)$ The space of bounded real-valued functions on a set S. 113

$C(X)$ All continuous, real-valued functions on a topological space X. 158

F_2 The free group on two generators. 128

$H \triangleleft G$ "H is a normal subgroup of G". 207

S^2 The unit sphere of $\mathbb{R}^3$ (the set of vectors in $\mathbb{R}^3$ with Euclidean norm $= 1$). 134

$\mathbb{B}^3$ The closed unit ball of $\mathbb{R}^3$ (the set of vectors in $\mathbb{R}^3$ with euclidean norm ≤ 1). $\mathbb{B}^N$ is the closed unit ball of $\mathbb{R}^N$. 140

$\mathbb{C}$ The set of complex numbers. 197

$\mathbb{C}^N$ N-dimensional (complex) Euclidean space. 12

$\mathbb{N}$ The set of positive integers. 127

$\mathbb{R}$ The set of real numbers. 3

$\mathbb{R}^N$ N-dimensional (real) Euclidean space. 6

$\mathbb{R}^N_+$ The set of vectors in $\mathbb{R}^N$, all of whose entries are non-negative. 11

$\mathbb{R}^S$ All functions on the set S with values in $\mathbb{R}$. More generally, Y^X denotes the collection of all functions on the set X with values in the set Y. 109

$\mathbb{Z}$ The set of integers. 117

$\emptyset$ The empty set. 110

$O(N)$ The collection of $N \times N$ orthogonal matrices. 200

$\omega(S)$ The weak-star topology on $\mathbb{R}^S$ (the product topology on $\prod_{s \in S} \mathbb{R}$). 109

$\prod_{s \in S} \mathbb{R}_s$ The topological product space of copies of $\mathbb{R}$ indexed by the set S. 110

$\mathscr{G}$ The group of isometries of $\mathbb{R}^3$. 140

$\mathscr{P}(S)$ The collection of subsets of a set S. 9

$\mathscr{R}$ The group of rotations of $\mathbb{R}^3$ about the origin. 134

$SO(2)$ The collection of 2×2 orthogonal matrices with determinant 1 ($SO(3)$ is the corresponding collection of 3×3 matrices). 145

Π_N The standard simplex in $\mathbb{R}^N$. 11

RBPM   "Regular Borel probability measure". 103

© Springer International Publishing Switzerland 2016
J.H. Shapiro, *A Fixed-Point Farrago*, Universitext,
DOI 10.1007/978-3-319-27978-7

Printed in the United States
By Bookmasters